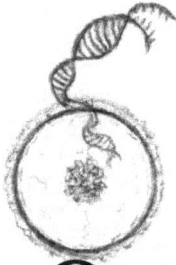

CELLULAR ORGANELLES

Legends of the Cell

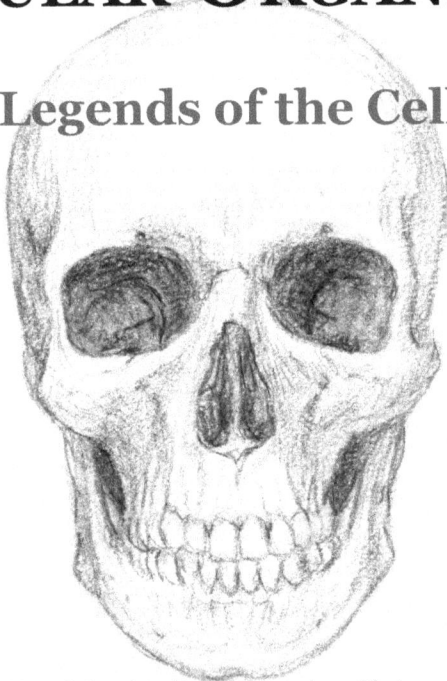

By William Brucker, Kate Schapira, Brenna Brucker, Christoph Schorl, Ava Lovato, Stacy Croteau, Carolina Veltri

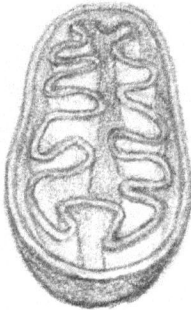

Illustrated by Grayson Armstrong, Cybele Collins, Andrew Peterson, Chris Boakye, Joao Paulo, Aziz Khoury

Edited by Melina Packer, Kate Schapira, Brenna Brucker

For information about permission to reproduce sections of this book, email:
providencealliance@gmail.com

Library of Congress Cataloging-in-Publication Data

Cellular Organelles: Legends of the Cell
Library of Congress Control Number: 2011963046
Brucker, William
ISBN: 0982818912

© 2012 Providence Alliance of Clinical Educators
IRC Section 501(c)(3) Public Charity
Brucker Books, Providence, RI

Printed in the United States

Acknowledgments

Peterson Foundation, Donald and Sylvia Robinson Family Foundation, Dr. Daithi Heffernan, Dr. Andrew H Stephen, John Luo, Melina Packer

Index of Concepts

FRAMING ANALOGY

Tension mounted in the post office of Rabbit Ridge, West Virginia as Tom Esterhaus, owner of "Tom's Tolerable Wines", confronted Bob Martin, the town's only postal worker, over a mismanaged shipment of rhubarb wine.

"Dang it, Bob, that woman from Riverside kept me on the phone over an hour yakking away and claiming I ruined her wedding 'cause she couldn't serve rhubarb wine to her guests. She said she would dismember me and everyone in town if she ever passed through here again, and quite frankly I believe her, she was a fierce looking woman. You can bet she asked for her money back too! I consider that you owe me that money, and I oughta point her your way too if she ever comes through here again. 'There's the man that wrecked your wedding,' I'll say."

"I'm so sorry, Tom. I think I figured out where things went wrong. That wine was addressed to 14 Main Street in Riverside. I forgot to put the ZIP code on the order, but I figured how many towns called Riverside would have a 14 Main Street? Who was to know that things would turn out like this? Live and learn, I guess. Incidentally, if you want that money from me you'll have to pry it from my cold dead hands."

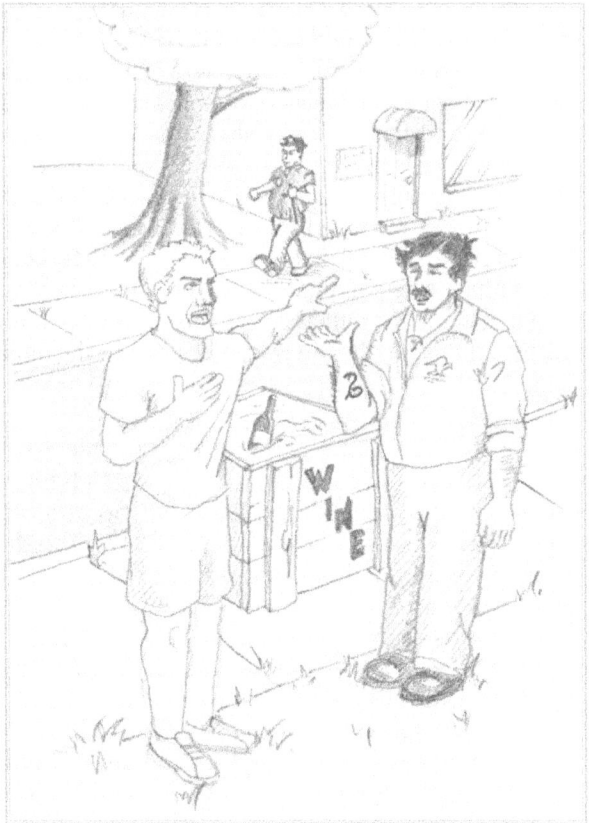

"That's a double bonus as far as I'm concerned. You ignorant goof, Riverside is the most common town name in America. 46 of the 50 states have one. That wine could be anywhere from Riverside, Maine to Riverside, Montana, but whichever state it is I won't be getting any orders from there again. This bonehead move of yours could put me out of business! What I really oughta do is force you to drink the next batch."

Without proper targeting information, shipments can never reach their destination. Street addresses and ZIP codes are the information that shipping companies use to determine where packages are supposed to go. The Golgi apparatus is the post office of the intracellular world, and molecular targeting sequences are the analog to ZIP codes and addresses.

Organelles are easy to think of as businesses staffed by protein workers. The proteins that work inside of the Golgi apparatus put targeting information onto proteins so that they can be shipped to the appropriate organelle or out of the cell entirely. Unfortunately, all it takes is one employee like Bob to screw up this targeting information and prevent proteins from being correctly delivered. There are always consequences—usually negative ones—when a shipment fails to reach its destination.

THE STORY: THE CONSEQUENCES OF GOLGI DYSFUNCTION

A 6-year-old girl is brought to your office because of difficulty moving. She has disproportionate facial features, particularly a broad enlargement of the forehead, jaw, and lips—it looks almost as if her face is too large for her head. Her nose, too, is on the wide and flat side, and her tongue is larger than normal. Her joints are unusually stiff and rigid and her spine has a prominent curve. Although she is six years old, she is closer to the size of a three-year-old child, and X-rays show multiple skeletal deformities. When you feel her abdomen you notice that her liver is much larger than normal. She exhibits multiple problems with her nervous system and mental functions. Blood tests reveal a high plasma concentration of enzymes that should be in lysosomes, not in the blood.

Scientific Connection

This young patient has Inclusion Cell Disease. Also known as I-Cell Disease, this genetic condition has no cure and typically causes death in childhood, usually from heart failure. The disease represents a critical failure in intracellular shipping that is directly linked to problems in the Golgi apparatus.

Lysosomal enzymes are proteins whose job is to break different molecules down so that their components can be recycled. If these molecules are not degraded (broken down), they accumulate inside cells, producing gigantic inclusion bodies in the cytoplasm. These gigantic inclusion bodies take disproportionate amounts of cellular space and interfere with the normal functioning of the cell; this in turn causes diseases in multiple systems, leading to the enlarged tissues that are responsible for the multiple deformities associated with this disease. The combination of stunted growth and multiple deformities of the face and bones are collectively known as gargoylism, a condition frequently associated with inclusion body disorders.

I-Cell Disease is caused by a defect in the protein whose job is to attach a phosphate to mannose residues on lysosomal enzymes. If the mannose residues are phosphorylated—have a phosphate molecule attached, like the ZIP code on a package—then the lysosomal enzymes will be correctly targeted to their destination: the lysososmes. In I-Cell Disease, the mannose residues of these proteins don't get phosphorylated, so instead of getting delivered to the lysosomes they get shipped outside of the cell. This is why the lysosomal enzymes show up at high levels in the blood and why inclusion bodies build up in the cells. Because the proteins never make it to the lysosomes, the lysosomes cannot do their job as well, and the substances that they should be breaking down pile up in the cell instead.

This article provides a case study and more in depth description of the disease: http://www.ncbi.nlm.nih.gov/pmc/articles/PMC2495859/pdf/postmedj00317-0077.pdf

⤳*Take Home Message*⤶
The Golgi apparatus is responsible for shipping proteins to their correct destinations. Targeting sequences on proteins are the cellular equivalent of zip codes and street addresses.

FRAMING ANALOGY-YOU GET WHAT YOU PAY FOR

Laurent Sanders smiled and whistled a tune as his Porsche sped down the long dirt road towards his new country estate in scenic Rabbit Ridge. He had one hand on the wheel and one arm around his new wife Delia, a beautiful woman more than 20 years his junior. Laurent is an extremely wealthy man who lives in the distant city of Riverside. He and his brother Charles own a string of businesses that are actually fronts for illegal enterprises. Laurent is driven more by the joy of cheating the system than amassing wealth. This love of a good swindle was why he was particularly excited to inform Delia not only that he had gotten a great deal on the price of the house but exactly how he had masterminded the whole thing.

Laurent shot Delia a wink. "I'm telling you baby, this country estate is going to just about knock you out. I can't think of a better way to spend a honeymoon than in the beautiful secluded countryside of Rabbit Ridge. The deal I got on this place is incredible! These Rabbit Ridge bumpkins are no match for a big city wheeler and dealer like me. The local construction company wanted to charge me 700 grand! What a joke! What would these goofs do with that money? They'd probably just buy some Armani overalls. I got the whole thing done for less than $50,000. Economic recessions are a wonderful thing. I found some farmers who were hard up for cash and I hired them as my construction crew for pennies on the dollar. Since I own a construction company back in Riverside I know all about building houses. I have been sending these guys letters once a week detailing exactly how I want things done and they send me pictures of their progress every now and then so I know things are going according to plan. Face it doll, you married a genius!"

Delia squirmed, underneath his arm, "If you were so smart you wouldn't have bought me that big iced coffee before a long car trip. I'm more interested in being with a rich man than a smart one, and I can tell you rich men's wives don't use the forest as a restroom!"

When they pulled into the driveway of their new vacation home it looked even better than the pictures had shown. The foreman was waiting for them at the front door and Laurent was overjoyed at the sight. Delia was less

impressed and just wanted to use the bathroom, the foreman told her that it was on the first floor and she quickly ran into the house. The foreman beamed with pride: "Well Boss, this is my best work ever. Heck, it's my only work ever. We got two master bedrooms, walk in closets, a great foyer, a giant living room, fireplaces, the best outhouses money could buy, and a spectacular well!" The color drained from Laurent's face as a furious scream erupted from the house and Delia burst through the front door in a fury, "I ran all over the house and there are no bathrooms anywhere in it!"

The foreman scratched his head and shrugged his shoulders, "What I meant to say was that you needed to go out the back door on the first floor and the outhouse is right there. I even built a trellis over it myself in honor of your nuptials. The roses growing through it are not only visually appealing but they are great at masking the smell. Consider it my personal gift to you."

Delia screamed and pulled her hair. "An outhouse! An outhouse! Laurent Sanders, I promise this is going to be a honeymoon you will never forget for the rest of your life... no matter how hard you try." She stormed into the house and slammed the door breaking the stained glass window embedded in it.

The foreman looked at a catatonic Laurent and whispered to him, "I don't know what she's so upset about. That deluxe outhouse you got is way better than the discount one I use." Laurent grabbed the foreman and slammed him up against the side of the house before snarling in his face, "Why didn't you follow my instructions and install the plumbing!"

The foreman winced and muttered, "We got all your other letters except for that one. We didn't know what to do so we improvised. It must have been Bob that screwed up. He's our idiot mailman. He's always losing letters and sending things to the wrong address. We followed any orders you gave us down to the last detail."

Laurent furiously hurled the foreman to the ground, "Your last order is to tell me where this mailman lives. When I find him I will tear him limb from limb. No one ruins Laurent Sanders' honeymoon and gets away with it!"

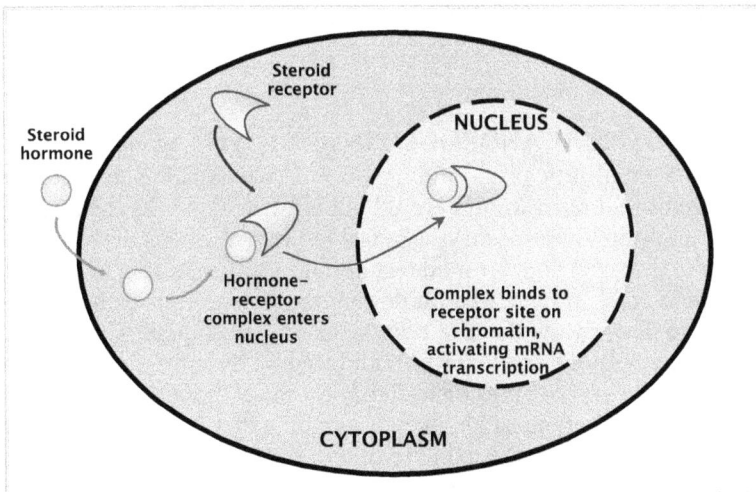

Scientific Connection

A source of major problems in the last story is that a critical message was not delivered and essential work was never completed. Because the mailman never delivered the message instructing the crew to install plumbing, the job was never done. If he had delivered the letter with the instructions then everything would have gone smoothly. The technical fault in this case lies with the mailman.

Messengers are also an important part of biology and have a similar role in the completion of tasks. The above story is conceptually related to the physiology of steroid hormones and how they interact with the nucleus in order to direct which jobs need to be done. The steroid hormone plays the role of the "letter" while the steroid receptor is the "mailman" and the nucleus is the "construction crew".

Steroid hormones interact with steroid hormone receptors in the cytoplasm. The hormone-receptor complex then travels to the nucleus where it activates key segments of DNA (genes) that represent the blue prints on how to build proteins designed to complete very specific tasks. Proteins are the workers of the biological world, but unlike the real world where workers are hired, in the realm of molecular biology they are made from instructions in DNA. The nucleus represents a library of information on how to make proteins to complete almost any job that you could imagine and even a few you probably couldn't.

Messengers like steroid hormones direct the hiring of proteins by telling the nucleus which jobs need to be done. If a protein in a cell senses an intracellular "fire," it sends a steroid messenger straight to the nucleus so that a protein "firefighter" can be made to put it out. However, if something in the sequence goes wrong and there is no steroid hormone receptor to deliver the steroid's message to the nucleus, then the requisite proteins never get made and the critical work never gets done. The following story is a real world example of what happens when a defective steroid hormone receptor prevents the delivery of a critical message to the nucleus and a very important job is never completed.

THE STORY-COMPLETE ANDROGEN INSENSITIVITY SYNDROME

Victoria Matthews is a beautiful 25-year-old woman. She has a slender physique and curves that most women would kill for. Her skin is as clean and clear as a glass of skim milk. Victoria works as a fashion model. She is never short of jobs and is actually the most popular model at her agency. Employers love that she is a little taller than most women. The photographers always request her because she never seems to develop any kind of body odor on long shoots. The lights used in the photography sessions are very warm and frequently leave models and photographers soaked in sweat. By the end of the day some of those studios can smell like a barn! Employers also request her heavily any time they have a job that requires a model to pose in a bikini. The waxing required for the removal of hair from bikini zones often leaves red marks that need to be airbrushed out. Victoria never seems to have this problem. Recently Victoria has been quite conflicted on whether or not she should have her testicles surgically removed.

Victoria comes from a very rural and medically underserved area and hardly ever visits a doctor. A few years ago she and her boyfriend were driving to a movie when they were involved in a car accident. When she was examined in the emergency room two strange solid masses were felt in her abdomen. Victoria said that she had them all her life and they had never given her trouble. She received only minor injuries but was given a series of imaging tests (X-ray/CT scan) to determine if she had any internal bleeding. The imaging tests revealed that not only was she missing all of her internal reproductive organs (no uterus, no ovaries) she actually had a pair of undescended testicles in her abdomen. Examination of her chromosomes revealed that although she had the appearance of a woman she was in fact a genetic man. She was not really bothered by the revelation and feels like a normal woman. Most days she doesn't give it a second thought. The thing that bothers her the most is that she will never be able to have a baby of her own but one day she hopes to adopt. She was told that it is possible for the undescended testicles to be a source of cancer so they should probably be removed. Victoria is most worried about the effect that surgical scars would have on her modeling career so she has been weighing the decision carefully.

Scientific Connection

This is a description of Complete Androgen Insensitivity Syndrome, a highly unusual condition that is a result of essential cells not getting a critical message. Within the first few weeks of life a developing human is neither male nor female. Proper cellular instructions are as vital to this period of human development as they were in the construction of Laurent's summer home. In the case of the summer home, instructions on how to install plumbing were never given so the construction crew did what they knew how to do and installed an outhouse and a well instead of indoor plumbing. When Victoria was a developing organism, messages were sent based on the instructions in her DNA to give her the external appearance of a boy. However, it is clear that a certain set of cells in her body never received this information. In the absence of specific instructions that direct the outward appearance to become that of a man, the default pattern of development is to create the outward appearance of a woman. This signal transmission error can be traced directly to the nucleus and its lack of response to cellular messengers known as androgens, a type of steroid hormone.

Victoria has a disease known as complete androgen insensitivity syndrome. Normally the androgen delivers a message to the recipient cells of the body which tells them among other things to take the shape of a man. The defect in this case lies with the androgen receptor (mailman), the protein designed to receive the androgen (letter) and carry it to the nucleus so that proteins can be made to carry out the orders of the message. In Victoria's case, her defective androgen receptors never interacted with the androgens produced in her body, the message never made its way to the nucleus, and proteins that would have carried out the critical orders to make her look like a man were never made.

Though externally Victoria appears to be a woman, internally she has male reproductive organs. The testicles develop in the abdomen independent of any steroid hormone messages before being pulled down into the scrotum by two little ropes known as gubernaculi. Since Victoria has the external reproductive organs of a woman and no scrotum for the testicles to descend into they remained in her abdomen. The lack of internal female reproductive organs (uterus, ovaries) is usually the reason that this disorder is identified before the affected individual is 20. A uterus is required for an individual to have a period. Usually if an individual has not had a period by age 15 a medical investigation is warranted. In Victoria's case, genetic studies would show an individual with a male genotype (46 XY) and a female phenotype. A major physiological consequence of this disorder is infertility, because while Victoria would have a vagina both the ovaries and uterus are required for pregnancy. Since Victoria has neither she can never have a baby. The psychological effects of the disorder are often severe but in this case Victoria is not significantly bothered and identifies herself as a woman.

In many ways the fact that the androgen "messages" were never delivered was a great advantage to Victoria in her modeling career. Puberty represents a time of androgen explosion. Many unpleasant aspects of puberty can be attributed to androgens. They increase the activity of glands in the skin that result in both acne and unpleasant body odor. Androgens also direct the growth of pubic hair under the armpits and around the genitals. Because Victoria had defective androgen receptors, these messages were never received by the nucleus and as a result she never had body odor, acne, or any pubic hair — all of which enhanced her career as a fashion model.

THE STORY-THE BEAUTY OF ANTAGONISM

Carolyn Walker is a 16-year-old girl with a terrible complexion. She developed a severe case of acne shortly after puberty and has become extremely self-conscious about it. After visiting her dermatologist she was prescribed a drug known as spironolactone to improve her complexion. Within a few weeks of taking spironolactone her acne began to clear up and her appearance began to improve.

Scientific Connection

Androgens deliver signals that increase the activity of sebaceous glands in the skin. Sebaceous glands produce oil that makes the skin smooth and luxurious. Highly active sebaceous glands are easily clogged and can quickly become infected and inflamed, giving rise to the pimples that are the hallmarks of acne. The volume of androgens secreted during puberty is greatly increased. As a result, the sebaceous gland activity goes through the roof, leading to an increased incidence of pimples during the teenage years. If the problem is that the sebaceous glands are overactive, then the solution is to shut them down. If the activity of the gland is directly related to signals delivered to the nucleus by androgens, then it is time to stop those messages from being delivered. The drug spironolactone interferes with both the creation of androgens and their effects on tissues like the sebaceous glands by blocking androgen receptors. By inducing a type of androgen insensitivity, spironolactone is able to decrease sebaceous gland activity and the severity of acne. While this is a good treatment for acne in women, it isn't such a good idea for young men because it can cause them to develop secondary sex characteristics associated with females (known as "feminization"). More on the role of spironolactone in acne can be found in this article: http://www.ncbi.nlm.nih.gov/pubmed/21413648.

⇝ *Take Home Message* ⇜

The nucleus controls cellular functioning by directing which proteins are made. Messengers like steroid hormones advise the nucleus on which jobs need to be completed and which proteins need to be made.

Cytoplasm: Murder in Rabbit Ridge

FRAMING ANALOGY

Rabbit Ridge is a small town with few notable events and even fewer people. The sight of two police cars parked outside of a house on 187 Orchard Street, and a pair of uniformed officers setting up crime scene tape, attracted a lot of attention from the locals. From the outside, the one-story ranch house looked like all its neighbors in the noontime light. Detective Klump and Sgt. Stubb stood on the stoop looking in.

"Something is definitely not right in there," Sgt. Stubb said after staring into the living room for a full minute. He shook his head slowly. "Nope, not right at all."

A small crowd of curious people was beginning to gather. Ms. Johnson, who lived at 183 Orchard Street, had called the police when she saw a puddle of blood seeping out from under the front door. Now she seemed about to explode with exasperation. "Not right?" she repeated. "There are human intestines all over the danged carpet! There's blood all over the furniture and a human heart on the table! There is a pair of kidneys nailed to that wall! Is 'not right' maybe a little bit of an understatement here?"

"Intestines," Sgt. Stubb said, shading his eyes and pointing to the carpet. "Those are the long ones, right?"

Ms. Johnson was jumping up and down with impatience. "Well?" she demanded. "Aren't you going to call the forensics people? Like on CSI? I watch a lot of television and I know how things are done!"

"They're on their way, Ma'am," Detective Klump told her. "But it's pretty obvious that someone got killed."

Ms. Johnson stared. "You mean you know who did it?"

Klump calmly replied, "No, Ma'am, but we know somebody did something and it's a fair guess to say that there has been a murder."

Sgt. Stubb coughed near Detective Klump's left shoulder. "But there's no body, Detective. How can we be sure someone's been killed?"

Klump turned toward his subordinate officer with a furious hiss. "I can see there's no body, Sergeant. These ... parts ... aren't supposed to be out here, they're supposed to be inside somebody. My years in law enforcement have taught me that when you see a person's insides on the outside that's never a good thing. I'm no doctor, but I know that much."

"Maybe it would help if I told you whose house this was," Ms. Johnson said sarcastically. "You know, like if you investigated."

"Whose house was it, Ma'am?"

"Bob Martin lived here."

"Bob the mailman?" Stubb turned to Klump. "Hey, you know what? Good news! I think I solved the mystery!"

"You mean you know who the killer is," Ms. Johnson almost shrieked, "and you're still not going to do anything?"

"Oh, I have no idea who the killer is," Sgt. Stubb assured her hastily, "but this might explain why no one's been getting any mail."

Scientific Connection

Internal organs are called "internal" for a reason. If you see blood and guts lying around, you can safely assume that whoever or whatever they came from is dead. The indirect evidence for violent death is extremely strong even if you can't see the body itself. This logic also applies to cells and is used clinically to determine organ damage and dysfunction. Organs are made of tissues and tissues are made of cells. If a large number of cells in an organ get killed then it is safe to say that the organ is not going to function as well as it used to, just as if you murder the only mailman in town then letters won't get delivered.

How do you know that an organ has been damaged without being able to see inside a person's body? The answer lies in the indirect evidence: the "blood" and "guts" of cells. When a cell dies, it lyses or "pops", sending its cytoplasm and proteins into the blood. If cytoplasmic proteins are present at high levels in the blood, you can assume that the cells they came from have been killed or seriously damaged.

Different tissues are made of different cell types, each of which has tissue-specific proteins in its cytoplasm. This means that sometimes you can determine which organ has been damaged based on which proteins you see in the blood. The following vignettes will explore the tissue-specific intracellular proteins that indicate when an organ has been damaged. By recognizing and identifying these clues, healthcare workers take on the role of "detectives", trying to make a diagnosis and therefore determine the identity of the "murderer" causing the disease.

⤙ *Take Home Message* ⤚

Cytoplasm is packed with proteins. When intracellular proteins are present at high concentrations in the blood it indicates cellular death and organ damage.

STORY 1: 75-YEAR-OLD WOMAN FOUND DOWN

The old woman lay at the foot of the stairs, her head bleeding. Shaking, her grandson crept closer. "Grandma," he whispered. She didn't move. Terrence Barrow felt like crying, but he remembered what his parents told him: In an emergency, you use that phone we gave you to dial 911, and then you call one of us.

Several hours later, while Terrence clutched her hand, Sarah Barrow said with exaggerated patience to the ER physician attending her, "I feel fine, young lady, and you claim the scans don't show any brain damage or concussion or whatever fool thing you can get from a bump on the head. You say I didn't have a stroke either. The last thing I remember is walking down the stairs. I must have just tripped and fallen—I understand we old ladies do that sometimes—but now that you've patched me up, Terrence and I have a movie to watch and some cookies to eat, and we'd like to go home."

The doctor shook her head firmly. "Not just yet, Mrs. Barrow. The tests show that you had a heart attack. You need to stay with us a little longer so we can check for complications and talk with you about preventing another attack."

"But I didn't have any chest pain or anything! How can you be so sure? What tests?"

"The blood we took while you were unconscious showed high levels of cardiac troponins—those are proteins found in the muscle cells of the heart, and they're supposed to stay inside of those cells and never be in the bloodstream."

Sarah sighed. "All right, Terrence, you better call your daddy again and tell him to bring over my things when he comes."

Scientific Connection

The most dangerous injury is the one that you miss and fail to treat. When someone is found unconscious, you cannot assume that they just tripped and fell. Possible causes of syncope (the medical term for fainting or passing out) have to be explored: it may be something like dehydration, hypoglycemia (low blood sugar—very dangerous), a seizure, stroke or heart attack. Not all heart attacks have the classic, "movie" symptoms of chest pain and difficulty breathing; many people have them and don't even know it. Sarah's head injury immediately drew attention because it was visible, and she assumed it was the only thing she had to worry about. If she had been released from the hospital without further exploration of her condition, she might never have known about her heart attack and easily could have suffered fatal complications from it in the near future. Because the heart attack was quickly identified, she and her doctors can take preventative action to minimize the risk of complications, and she and Terrence will have more afternoons together.

The standard series of tests to explore the possible causes of syncope are known as a "syncope work-up". One test in the work-up is to examine the blood level of cardiac troponins. These proteins are specific to the cytoplasm of cardiac myocytes, the cells that make up the muscle tissue of the heart. When the blood vessels that supply the heart become clogged with cholesterol deposits, it prevents these cells from getting enough blood (with its cargo of nutrients and oxygen). These cells cannot make enough energy to survive and their deaths result in the release of cytoplasm and cardiac troponins into the blood. A high level of cardiac troponins in the blood means that a lot of heart muscle cells have been killed—otherwise known as a myocardial infarction (heart attack).

⇀ *Take Home Message* ↽
Testing for levels of cardiac troponins in the blood is one way to determine if the heart has been damaged.

STORY 2: EXTRA STRENGTH LIVER DAMAGE

Dante Correia is 27 and works as a corrections officer. As a result of stress at work, he suffers from frequent headaches. Two days ago, Dante bought Extra-Strength Tylenol on his way to the prison because he's heard it's effective. His headache was so bad that he couldn't concentrate, couldn't catch the little cues that might mean trouble. He couldn't go home—he needed the hours, and he didn't want to leave the other COs shorthanded. Dante read the bottle and scoffed at the recommended dose. Two pills? Maybe for a regular guy but no way that would be enough for him. He took 18.

Shortly after taking this massive amount of medication, he began to vomit and they sent him home anyway. Now two days have passed. He's become increasingly weak and confused and has terrible pain on the right side of his upper abdomen. The whites of his eyes have a yellow tone, as does his skin. He goes to the hospital.

Dante's blood tests show elevated levels of two proteins, AST and ALT. The physician who sees him tells him that an overdose of acetaminophen damaged his liver—maybe beyond repair. If Dante is lucky his liver will get better over the next few days. If his liver continues to fail, and he doesn't get—and survive—a liver transplant, he will die.

Scientific Connection

In recommended doses, acetaminophen is a safe drug and an effective pain reliever. In excessive doses, it will destroy your liver. It is important to remember that the dose makes the poison: anything can be toxic in high enough quantities. Drug manufacturers make every effort to inform you (on the side of medication bottles) of the appropriate amount of a drug to take. As Dante's case shows, these recommendations should be taken seriously.

Extra-Strength Tylenol pills each contain half a gram of acetamino-phen. The therapeutic dose is 1 gram and each pill contains half a gram. The recommended upper limit is 4 grams (8 pills) a day. 6-7 grams is the estimated amount required to damage the liver. The 18 pills Dante ingested contained 9 grams of acetaminophen—more than enough to wreck his liver. The yellow tinge to his skin came from jaundice. Jaundice is the visible result of high levels of bilirubin in the blood. Bilirubin is a small molecule that the body constantly produces as a waste product; it is detoxified and eliminated by the liver. Jaundice is often a sign that the liver is heavily damaged and can't do its job.

High doses of acetaminophen are toxic to hepatocytes, the cells that make up liver tissue. When hepatocytes are destroyed they release intracellular proteins like AST and ALT, which are considered markers of liver damage. Blood levels of AST and ALT are used to track the progress of the liver's recovery. If the levels increase, more cells have died and the damage is worsening; if they decrease, the liver might be getting better.

➝ Take Home Message ➝

In sufficient doses acetaminophen is toxic to the liver. When liver cells are destroyed they release the proteins AST and ALT, which are used to detect liver damage.

STORY 3: YOU CAN'T BELIEVE EVERYTHING YOU READ (ONLINE)

Philip Tavinoff slouched down the Rite-Aid aisle with a 300-mL bottle of Ny-quil-D in hand. He had a cover story ready about his mom's terrible cough, but the clerk took his money and handed him his bag without even looking up. At home, he switched on his laptop and looked at the page again: "Dextromethorphan is an opi-ate, like morphine." The Wikipedia entry assured him that, like morphine, it would get him high, although since it wasn't as strong as morphine he would need to take more. In the screen's glow, he read the label on the box and saw dextromethorphan on the list of active ingredients. No need to read any further: he opened the bottle and swigged the contents, wincing at the taste. "Better safe than sorry," he thought, and chugged the rest.

48 hours later, Philip was definitely sorry and not particularly safe. His upper right abdomen hurt horribly, even after throwing up several times, but it wasn't until his skin and eyes began to turn yellow that his mother took him to the emergency room, where blood tests revealed high levels of AST and ALT. As his condition worsened, the doctor told his mother that since he was a minor, they might be able to move him up on the liver transplant list, but noted privately that this might be natural selection in action.

Scientific Connection

Philip did his homework but still got an "F" in amateur pharmacology. Nyquil is not pure dextromethorphan (which is given to stop you from coughing, not to get you high). Nyquil is actually a mix of several drugs that are designed to prevent cough, reduce inflammation and pain, and help you sleep. Taken at recommended doses it is a safe and effective formulation.

But one of the drugs in the cocktail is acetaminophen, which can damage the liver in large doses. Every 30 mL of Nyquil D contains 1 gram of acetaminophen. If Philip drank 300 mL of Nyquil then he ingested 10 grams of acetaminophen; it only takes 6-7 to seriously damage the liver. High doses of acetaminophen are toxic to hepatocytes, the cells that make up liver tissue. This is why his blood levels of AST and ALT are highly elevated and he has the same signs of liver failure as poor Dante. Internet sources that recommend the misuse of pharmacological agents to get high are very dangerous. The only people more foolish than those who provide that information—often, as in this case, misinformation—are the fools who make use of it. For a pertinent look at ignorance in action, see either of these links; just looking won't injure your brain cells that much.

http://answers.yahoo.com/question/index?qid=20090821195902AAD5AvH
http://answers.yahoo.com/question/index?qid=20090315184847AA2Hx9o

⇝ *Take Home Message* ⇜

In sufficient doses acetaminophen is toxic to the liver. When liver cells are destroyed they release the proteins AST and ALT, which are used to detect liver damage. Many over-the-counter drugs contain multiple chemicals, and misusing them can result in severe and unintended consequences.

STORY 4: REASONS NOT TO STEAL YOUR GRANDMA'S MEDICATION

Jessica Bigelow nudged her 16-year-old twin sons and muttered, "Ask your grandma how she's feeling."

"How are you feeling, Grandma?" Marvin and Mitchell asked in chorus. They glared at each other, but Grandma Audrey, recovering from a car accident, didn't seem to notice anything. "Did they give you anything, Ma?" Jessica asked. "You seem a little out of it."

"They gave me a prescription for something to take as needed—every four hours or so. I took one, but I don't like it. Vicodin, I think it's called." The two boys looked at each other, not a glare this time, and Marvin said quickly, "So, Grandma, have the ladies from church been by yet?" Pleased by the attention from her usually surly grandson, Audrey began describing the parade of visitors she'd had while

Jessica looked on approvingly. Neither woman noticed Mitchell's departure from the room, or his return with one hand in the kangaroo pocket of his hoodie. Later, the boys divided up the contents of the Vicodin bottle. "This is supposed to be amazing," Mitchell declared. "Like they could cut your legs off right in front of you and you wouldn't even care."

By the next morning, the boys could no longer hide that something had gone horribly wrong with their plan. Jessica took them to the emergency room, where they refused to admit that they had taken their grandmother's medication, but where a doctor—seeing their yellowed skin and eyes, palpating the painful upper right area of their abdomens, and noting their weakness and disorientation—immediately tested their blood for elevated levels of AST and ALT. Confirmed in his suspicions, he put the boys on the list for liver transplants but counseled their mother to prepare for the worst.

Scientific Connection

Vicodin is actually two medications combined into a single pill: hydrocodone and acetaminophen. Hydrocodone is an opiate that is used to reduce pain. It gets public attention because it can provide users with a high and because it can be addictive, but in fact the body can tolerate a lot more of it at once than Vicodin's other, less famous component, acetaminophen. The amount of acetaminophen in Vicodin changes with the prescription, but it can be as much as 1 gram per pill.

The therapeutic dose of acetaminophen is 1 gram; 6-7 grams is the estimated amount required to damage the liver. Taking Vicodin to get high—that is, taking more pills than the prescribed dose—can easily result in acetaminophen-mediated liver damage, and has done so in many cases of Vicodin abuse. Studies show that few people know that Vicodin contains acetaminophen or that acetaminophen is dangerous in large quantities. Percocet is another opiate/acetaminophen combination.

High doses of acetaminophen are toxic to hepatocytes, the cells that make up liver tissue. When hepatocytes are destroyed they release intracellular proteins like AST and ALT, which are markers of liver damage. Blood levels of AST and ALT are used to track the progress of the liver's recovery. If they decrease, the liver might be getting better; if the levels increase, more cells have died and the damage is worsening.

⇝ *Take Home Message* ⇜

In sufficient doses acetaminophen is toxic to the liver. When liver cells are destroyed they release the proteins AST and ALT, which are used to detect liver damage. Because Vicodin is a combination of hydrocodone and acetaminophen, its abuse can easily result in liver damage.

STORY 5: ALCOHOLIC LIVER DAMAGE

A 56-year-old man is brought into the emergency room by his daughter. He displays classic signs of acute liver failure (he has a fever, is confused, weak, nauseated, has terrible pain in the right upper quadrant of his abdomen and a yellowing of his skin and eyes). His daughter tells the emergency room physician that her father has been drinking alcohol heavily for nearly 20 years since his wife passed away. Blood tests show elevated levels of AST and ALT.

Scientific Connection

The processing of alcohol by the liver generates toxic substances that damage and destroy liver cells. Chronic consumption of high levels of alcohol can seriously damage the liver, resulting in elevated levels of AST and ALT in the blood. This is a case of alcohol-induced inflammation of the liver—known as alcoholic hepatitis—that is a result of decades of alcohol abuse. The liver is pretty tough and can handle alcohol, but if it gets pushed over the edge, as in this case, alcoholic hepatitis can easily lead to liver failure and a horrible death. Certain events can precipitate alcohol-induced hepatitis, such as drinking heavily while taking anything with acetaminophen. Also, any present infectious diseases of the liver (like viral hepatitis) combined with alcohol use equals massive hepatocyte death and huge AST/ALT increases in the blood. You can think of a viral hepatitis as a slow smoldering fire that is burning up your liver. Drinking alcohol while you have viral hepatitis is like pouring gasoline (or Bacardi 151) on that fire and destroying it many times faster than the viral disease would do alone.

→ Take Home Message ←
Repeated alcohol abuse will result in serious liver damage. Alcohol abuse combined with acetaminophen use or viral hepatitis will increase the extent and speed of that damage.

STORY 6: LAST MAN STANDING
You have been following the trends in blood levels of AST and ALT in a man diagnosed with liver failure. You find that over the last week, the levels have actually been dropping! But his upper-right-abdominal pain, his jaundice, his weakness and disorientation, and his vomiting and diarrhea have all been increasing. Is he recovering or getting worse?

Scientific Connection
The patient is getting a lot worse and will likely need a liver transplant if he is to survive. The fact that the levels of AST and ALT are lower may seem positive, but that they are dropping while the physical signs of liver failure are getting worse indicates that there are not many hepatocytes left to be destroyed. The dead cells have already released their AST and ALT; fewer dying cells means less AST and ALT will accumulate in the blood.

→ Take Home Message ←
Decreasing levels of AST and ALT in the face of worsening physical signs of liver failure means that the liver is mostly destroyed.

STORY 7: COLLATERAL DAMAGE
You meet with a 57-year-old patient for a follow-up visit. A few weeks ago, his blood tested high for cholesterol, which can lead to heart disease; you prescribed medication to lower his cholesterol. Your tests today are to see if the medication

is working. When you tell him that his cholesterol levels are lower but that he will need to switch to a different medication, he is confused and demands an explanation. The blood tests that revealed a lowered cholesterol level also revealed tremendously elevated levels of AST and ALT; if he keeps taking the first medication, you explain, he could damage his liver permanently.

Scientific Connection:

When you take a prescription drug, it doesn't affect only the part of the body it's prescribed for and leave the rest of the body alone; the drug doesn't "know" what needs to be "fixed". A drug is a foreign chemical that usually affects multiple organs to different extents. Many drugs are toxic in higher doses and therapeutic in lower doses, and the chemical process that helps one part or function of the body may harm another. This potential harm or collateral damage is popularly known as a "side effect," often listed in tiny print at the bottom of ads for medications.

Pharmaceutical therapy is always a cost vs. benefit strategy because collateral damage frequently occurs, as it did in this patient's case. The drug that his physician prescribed had the desired effect of lowering his cholesterol, but the side effect of destroying his liver cells, causing them to release AST and ALT into the blood. There are numerous drugs that require AST and ALT levels to be monitored in case they damage the liver. (You can see some of them in this article: http://www.factsandcomparisons.com/assets/hospitalpharm/HepSp4-01.pdf.) If this patient's doctor had only looked at whether the drug was doing its intended job (lowering his cholesterol) without checking to see if it was doing anything else (damaging liver cells, for example) the drug that was supposed to help him in one way could have severely harmed him in another. One of the goals of pharmaceutical research is to maximize the benefits of new drugs while minimizing their potential risks.

⇢ *Take Home Message* ⇠

Drugs frequently have adverse effects that should be weighed against the benefits of therapy. Many drugs can cause liver damage and blood levels of AST and ALT are monitored as a way to detect it.

STORY 8: ESCAPE FROM IRAQ

1997, New York City.
Notes in hospital file for Sadiq Ibrahim, male, age 26.
Patient admitted with high fever, vomiting, diarrhea, pain in upper right quadrant of abdomen. Yellowed skin and eyes (jaundice). Blood tests show highly elevated levels of AST and ALT.

When conscious, patient seems agitated. He was reluctant to give his name or to stay in the hospital. The following is a transcription of a statement made by patient.

"I don't mind this pain so much, Doctor. Only three weeks ago I was enduring much worse. You wouldn't think, looking at me now, that I was an athlete? I was on the Iraqi national soccer team, and when we won, everything was great. Big heroes. But if we lost, soldiers took us to the offices of the Iraqi National Olympic Committee. Doesn't sound so bad, right? But the president of this Committee was Uday Hussein, the dictator's son, and this 'office' was a prison and a torture chamber. His men tied my team captains to Jeeps and dragged them across concrete—they beat us with rubber hoses while we ran around a track. They pointed guns at us and ordered us to dive into a swimming pool and swim across it, but this swimming pool was filled with sewage—human waste. I can assure you it was not fresh either, we could smell it a mile away. If we didn't dive in headfirst they whipped us and made us jump again. At least this pain, I feel in a clean bed far away from that nightmare."

Scientific Connection

There's a reason you're supposed to wash your hands after using the bathroom. People who contract hepatitis A—one form of viral hepatitis—often do so by getting fecal matter in their mouths (known as fecal-oral transmission). Hepatitis A is one of the most common causes of liver disease and fecal-oral transmission makes it highly contagious and prevalent. In the majority of cases the body can defeat it and heal, but it does sometimes do enough damage to cause liver failure. Many states have begun vaccination campaigns to reduce hepatitis A infection.

When the soldiers forced Sadiq Ibrahim to dive headfirst into sewage he almost assuredly swallowed some and with it the hepatitis A virus. Hepatitis A has a three-week incubation period (between exposure and the development of symptoms), so Mr. Ibrahim did not begin to suffer from symptoms until he had already made his escape to the U.S.

⇀ Take Home Message ⇀

Hepatitis A is a viral infection of the liver that can cause serious damage. It produces the set of symptoms consistent with damaged liver cells. It is spread mainly by fecal-oral transmission, making handwashing an important part of reducing this disease.

STORY 9: A DIRTY STORY

You're the new doctor in a town so small it only has one restaurant. In your office today is 20-year-old Bradley, who has a high temperature. He tells you that he's been throwing up and "having the runs" for several days, and that he's had a stomachache "here"—pointing to the upper right quadrant of his abdomen. His skin and the whites of his eyes have become yellowish during that time, and when tested, his blood shows elevated levels of AST and ALT. You ask him if he has taken or is taking any medication; he says no. You ask if he has eaten anything recently that could have been contaminated, or that made him feel sick at the time. "Now that you mention it," he says, "I went to the Kozy Kitchen a couple days ago and I was sick to my stomach the next day. I guess it's kind of a dirty place. I heard one of the guys in the kitchen yell to someone, 'Get that donkey out of there and clean up its mess!' but I thought he was kidding. Then there was a noise like someone trying to push a big animal out the back door."

"A donkey?"

"Yeah, then he said something like, 'No time to wash your hands, get back to work.' But he must have been kidding, right? Aren't there inspectors and stuff for that?"

After prescribing a course of treatment for Bradley, you look up the number of the state Board of Health.

Scientific Connection

This is another case of hepatitis A. Fecal matter contaminating food is a frequent cause of hepatitis A outbreaks. It can come from food service workers who don't wash their hands after using the restroom; from poorly washed vegetables and fruits that might be contaminated with fecal matter before coming to the restaurant; or from the feces of animals, such as donkeys (but more commonly rats or mice), in the kitchen. The larger a supplier is, and the further a food product has to travel, the more chances it has to become contaminated. In 2003, a major outbreak in Pennsylvania was traced to unwashed contaminated vegetables used as a garnish in a Chi-Chi's restaurant. You can read more about it here: http://www.billmarler.com/key_case/chi-chis-hepatitis-a-outbreak/.

Bradley is right: we do have "inspectors and stuff for that." Call your state's Board of Health if you see something that makes you suspect that a restaurant is not meeting sanitation standards: you can reduce a major health hazard to the community and prevent your fellow diners from suffering as Bradley has.

⇀ Take Home Message ⇀

Fecal contamination of food can cause outbreaks of Hepatitis A. Sanitation in the food service industry is essential, both before the food is prepared and when it is prepared.

STORY 10: DANGERS OF FINE DINING

It's summer in Florida, and heavy rains have caused significant flooding in many areas. Today in the ER you are attempting to get some answers from Laura Perry, attorney-at-law, 40 years of age, about what might have caused her symptoms of liver damage: upper-right-abdominal pain, vomiting, diarrhea, fever and progressive yellowing of skin and eyes. Blood tests show highly elevated levels of AST and ALT. When you ask if she might have eaten contaminated food, she becomes indignant: "I don't eat in places like that! The last restaurant I ate in was Le Poisson Enchainé, when our firm went out to celebrate a successful verdict."

"They're a seafood restaurant, aren't they?"

"Yes, and they're famous for their oysters in particular. Fresh and delicious—and expensive. I'm sure there couldn't have been anything wrong there." But as she says it, her expression changes. "Raw oysters?" you ask.

"Yes, and ... I've just remembered, the two partners who went with me had to call in sick and cancel appointments with clients recently. I haven't seen them since. Could it have been the oysters? I would never have expected it from such a renowned restaurant."

Scientific Connection

If you guessed that this is another case of hepatitis A, you're right. Everyone who ate the oysters is now sick so it's clear that they were the cause. But where did the fecal matter come from? The restaurant deals in raw oysters frequently and has a good reputation, so its handling of the oysters is less likely to be the source of fecal contamination.

The key to this mystery is the series of rainstorms and resulting floods. Conditions that cause flooding can cause the sewage system to overflow and enter other waterways like rivers and bays, resulting in their contamination. Oysters are filter feeders: they suck in water, retain any particles or small organisms, and let the rest of the water out. This filtering allows them to concentrate material in the water. If oysters are in water contaminated with fecal matter, they will concentrate it within them, making them reservoirs for food-borne illnesses like hepatitis A.

Cooking reduces, but does not eliminate, the chances of contracting this virus. Hepatitis A is frequently associated with uncooked oysters and shellfish in general, so eat these foods with caution, especially if there has recently been a flood in the area where they come from; such an event is discussed in this article http://www.chattanoogan.com/articles/article_73351.asp.

⇢ Take Home Message ⇠

Filter feeders like oysters can accumulate toxic substances if they are in contaminated waters. Consuming uncooked shellfish can put you at risk for hepatitis A infection.

STORY 11: MUSCLE CELL DESTRUCTION: RATTLESNAKE ATTACK

You are a transplant surgeon having a preliminary meeting with Victor Vargas, who has been on dialysis for several years while waiting for a kidney transplant. A donor who is a good match has finally turned up. When you meet Victor for the first time, you are surprised to see that he has a prosthetic leg, and ask him if the loss of his leg is connected in any way to the problems with his kidneys.

Victor tells you that five years ago, he was hiking alone in Brazil when he accidentally startled a rattlesnake. He panicked and tripped, frightening the snake further; it bit him on the thigh as he was trying to crawl away. The nearest hospital was miles away, and by the time he was able to hobble there, it was almost three days later. He was in terrible pain; his wounded leg had swollen to nearly double its normal size and was almost black in color. Doctors gave him treatment to minimize the effects of the venom, but his kidneys had already begun to fail—due to the extremely high levels of the protein myoglobin in his blood, the result of muscle cell destruction caused by rattlesnake venom. He is looking forward, he says, to spending less time hooked up to the dialysis machine, and maybe one day hiking again.

Scientific Connection

Myoglobin is a protein inside the cytoplasm of muscle cells. As long as myoglobin stays inside the muscle cell, it performs its function and doesn't have any negative effects on the rest of the body. When muscle cells are destroyed, myoglobin makes its way into the blood and if it reaches high enough levels, the kidneys can be damaged or destroyed.

Rattlesnake venom is a mix of chemicals that harm the body in a variety of ways. One nasty effect of this venom is to destroy muscle cells in the area of the bite. A bite to the calf muscle from the Brazilian rattlesnake can destroy enough muscle cells and release enough myoglobin to cause serious kidney damage within 24 hours. The thigh is much larger than the calf (more muscle cells) and a bite there could result in the release of much more myoglobin. After three days without treatment, it is no surprise that Mr. Vargas's kidneys were destroyed.

The kidney is a vital organ that has roles in maintaining almost every other organ, so if it is destroyed the rest of your body will soon follow. One of the kidney's main functions is the filtering of wastes produced within the body. Dialysis is an imperfect replacement for a kidney, filtering wastes from the blood as your kidneys would, and can keep you alive until you can get a transplant.

Dead and destroyed muscle tissue has to be cut out of the body; otherwise, it is a breeding ground for infection. So many muscle cells were destroyed by the venom that the bitten leg had to be amputated. This article is a case study of envenomation by the Brazilian rattlesnake and resultant muscle tissue destruction and kidney failure: http://www.sciencedirect.com/science/article/pii/0041010185903678.

⇢*Take Home Message*⇠
Destruction of muscle cells releases myoglobin, which in sufficient concentrations can destroy the kidneys. Kidney failure is likely to follow whenever a large number of muscle cells are damaged.

STORY 12: ELECTRICAL BURNS: DEEPER THAN YOU THINK

You are an emergency room physician seeing a young man who has a severe electrical burn on his leg. His name is David Sanderson and he is 16 years old. He fell in with the wrong crowd and decided to join a gang. The local gang of tough kids said that he could become a full member if he could graffiti their insignia in 25 different places in town. One place that he chose to tag was an old electrical substation. He wasn't there for long before he was struck by a gigantic arc of electricity in the leg. He limped away from the substation, in some pain but otherwise feeling fine. David thought he could tough it out for a day or so, but when the leg began to swell up, he decided to come to the hospital. He has a blackened scorch mark on his thigh where the electrical arc hit him and another one on the bottom of his foot where the electricity left. Aside from the scorch marks and the swelling, there doesn't seem to be any visible damage to the leg from the burn. His blood demonstrates extremely high levels of myoglobin.

Scientific Connection

High voltage electrical burns are extremely dangerous because much of the damage is hidden from sight. Electricity is like an invisible knife that enters one side and leaves on another, piercing and destroying everything in between. These burns are much more dangerous than they look because a lot of the damage is internal and muscle tissue is highly susceptible to destruction. If muscle cells are destroyed they release myoglobin, which proceeds to destroy the kidneys if it accumulates in high enough levels. The electrical arc pierced David's thigh and exited at the bottom of his foot, destroying muscle tissue and a huge number of muscle cells along its course. The damage resulted in the release of massive amounts of myoglobin and the likely failure of his kidneys. Electrical service stations are extremely dangerous places and should only be entered by professionals. This is based on a real story in which the graffiti artist was not so lucky: http://www.trainorders.com/discussion/read.php?1,1409288. Because of their hidden nature, all electrical burns warrant medical attention to prevent the accumulation of myoglobin in the blood and the destruction of the kidneys.

⇀ *Take Home Message* ↽

High voltage electrical burns are usually much more serious than they appear and need immediate medical attention. Electrical burns can result in the destruction of muscle cells, the release of myoglobin, and failure of the kidneys.

STORY 13: ELECTRICAL BURNS: BODY OF EVIDENCE

You're an ER physician in a small town. There's not much to do here, and your main patients in the emergency room are high school kids whose attempts to entertain themselves have caught up with them. Your newest patient, Brandon Miller, is brought to the emergency room by his friends. Apparently they were playing a game to see who could climb to the highest point in the town's electrical substation. Brandon climbed the highest but then screamed and fell from the tower hitting the ground with a sickening thud. He has been unconscious since the fall, but imaging studies do not show any damage to the brain or the rest of his internal organs, and it seems like he only sustained minor cuts and bruises from the fall. You remember an instructor in medical school explaining that the most dangerous injury is the one that you miss. With that in mind, you search the rest of his body and find a scorch mark on his thigh and another one on the bottom of his foot.

Scientific Connection

Brandon was climbing the tower and when his friend weren't looking was hit by a bolt of electricity and fell off the top of the tower. Everyone was focused on the damage from the fall and missed the electrical burn. Since electrical burns destroy most of the tissue that they pass through many muscle cells were likely destroyed. Because the electrical burn was discovered quickly, measures could be taken so that myoglobin did not destroy Brandon's kidneys.

⟿ *Take Home Message* ⟿

High voltage electrical burns are usually much more serious than they appear and need immediate medical attention. The most dangerous injury is the one you miss and fail to treat.

STORY 14: CRUSHING INJURIES

Kurihara, Japan, 2011.

As soon as the earthquake's aftershocks faded, the rescue teams moved in to look for survivors, working until they were exhausted and others took their place. But there were many injured people needing immediate attention and thousands of pounds of rubble to move. It was nearly six hours before rescue workers were able to free Keiko Asusada from the brick wall that had crushed both her legs, cutting off all sensation and motion.

Blood tests would have revealed high levels of myoglobin in Mrs. Asusada's blood if anyone had been able to do them. The hospitals were full of thousands of injured survivors of the earthquake and tsunami. The rescuers hoped that they could get her adequate medical attention before her kidneys were destroyed by the myoglobin.

Scientific Connection

You can probably guess by now that muscle cells have been destroyed and myoglobin has leaked into the blood. The muscles have been crushed by a massive amount of rubble causing a "crush injury". The impact alone can destroy some muscle cells, but more serious is that the weight compresses the blood supply to the crushed tissue. Muscles of the arm and leg can go without blood for around 4 hours, but after that they start to die and release their cytoplasmic contents (like myoglobin) into the surrounding blood. Due to the massive weight compressing the limb blood can't flow in or out. While the blood in the trapped legs is accumulating toxic substances it can't reach the rest of the body because the debris acts like a giant tourniquet. The trouble begins when the debris is removed: the blood, now loaded with myoglobin and other toxic substances, can come back to the rest of the body and damage multiple organs, including the kidneys. For an article on crush injuries and the complications that can result from them: http://mediccom.org/public/tadmat/training/NDMS/crush.pdf.

⤳ Take Home Message ⤶

Crush injuries can result in the destruction of muscle tissue and dangerous levels of myoglobin in the blood, potentially leading to kidney failure. The other toxic intermediates are often more lethal in the short term.

FRAMING ANALOGY: TACTICAL LABOR INHIBITION

During World War I, Germany was fighting a two-front war, with most of Western Europe attacking them from one side and Russia attacking from the other. To reduce the power of at least one of their enemies, Germany created a devious strategy to take Russia out of the equation.

German leaders recognized that Russia was in a state of social turmoil and poised for economic collapse. The Russian economy depended on a large group of people working for low wages. Factory workers and other laborers in Russia were treated unfairly by their wealthier bosses, worked in harsh and dangerous conditions, and had low status and little power in making decisions in the country. Because of this, the potential for labor organizers to cause massive strikes were a major concern for the Russian government: Russian workers wanted to take power into their own hands, and they were ready to listen to anyone who encouraged them. To try to prevent this, the Russian government exiled many prominent Russian labor organizers, among them Vladimir Lenin, whose name you may recognize.

To capitalize on this turmoil and weaken their enemy, German leaders drew up a plan with Alexander Parvus (like Lenin, a Marxist revolutionary). The goal of this plan was to cause widespread labor strikes in Russia, known as "labor inhibition". These, in turn, would result in the country's economic paralysis and force its withdrawal from the war. The plan arranged for Germany to fund Russian Marxist revolutionaries who would act as labor agitators and precipitate strikes all across the country. To increase the success of this plan, the Germans smuggled Lenin back into Russia on a sealed diplomatic train. The conspirators hoped that unleashing Lenin on a vulnerable Russia would cause massive social upheaval and widespread strikes.

Even before the plan was enacted, many war-weary and frustrated Russian factory workers had already begun to strike. The German tactics bore fruit as massive and widespread strikes, accompanied by intense social unrest and the voicing of long-suppressed resentments, effectively shut Russia down. Russia was economically and socially destabilized by these events and had to withdraw from the war. Labor agitation with German sponsorship fanned the fires of economic and social dissatisfaction into a conflagration known as the Bolshevik Revolution, which ended with the murder and/or exile of many of Russia's former leaders, and led to the Communist Party's takeover of Russia and the establishment of the Union of Soviet Socialist Republics (now dismantled). Germany, however, ended up losing the war.

Scientific Connection

When groups of workers band together and forcibly prevent productivity it is known as a strike. Strikes can deal financial deathblows to businesses and are neither initiated nor responded to lightly. If one company's employees go on strike, their nation's economy might suffer a little, but if every major industry were to go on strike at once there would be far-ranging social, political, and economic consequences, as in the above example.

Like industries, biological systems must continuously do work and maintain order if they want to survive. The production of ATP is essential because it provides energy. Energy is the potential to do work, but potential is only valuable if it can be realized. A factory without workers to make its product is in as weak a position as workers who have no one to compensate them for their labor. Possessing sufficient energy to complete a task is as important as having the means to carry it out.

Proteins are the workers of the biological world. They are responsible for maintaining the health and integrity of the cell as well as accomplishing the cell's other goals or functions. These tiny and seldom appreciated laborers are the base on which biological systems are built and their constant toiling allows cells, tissues, organs, and organ systems to stay viable. Any substance that prevents proteins from being made or distracts them from their duties is usually phenomenally toxic. Inhibiting the activity of proteins on a wide scale will almost assuredly result in cellular death. Microbes are locked in a constant struggle for survival; they fight hard, play dirty, and in their own way are more devious than any human. The inhibition of cellular work by interfering with protein synthesis is a commonly used tactic that is both effective and deadly. When bacterial toxins interfere with protein synthesis in human cells the result is disease. Humans have caught on to this trick and utilize drugs that selectively disrupt bacterial protein synthesis to fight infections.

⇝ *Take Home Message* ⇜

Proteins are the workers of the biological world. A substance that inhibits widespread protein synthesis or protein function is typically extremely deadly.

STORY 1: THE EPIDEMIC- JANUARY, 1925.

Winter in Alaska makes the winters of New England seem like spring, thought Dr. Lawrence Harberg, physician for the rural town of Nome. Travel in and out of the town was impossible in the record levels of snow, ice and cold. This was his first winter here and he was afraid it might be his last winter anywhere. "Where is Annie?" he asked his nurse. "I told her mother to bring her back in today if that sore throat didn't get better—what's wrong?"

"Annie died two days ago," Sister Mary Katherine told him with tears in her eyes. Dr. Harberg snapped, "Don't get emotional now, Nurse, I won't have it," but he was badly shaken. The fifth of his young patients to die in a week—but surely sore throats were normal in the winter? Sister Mary Katherine blew her nose and reminded him, "Doctor, you have a patient waiting."

"Well, well," he said when he saw the two Yupik girls, "which one of you is my patient?"

"Her," the older girl said, pointing at her sister. "She doesn't speak English. Her throat hurts and it's hard for her to eat. And she says she's hot. She keeps throwing the blankets off."

Cold as the weather was, a chill went down Dr. Harberg's spine. "Well, let's have a look," he said, and gestured for the little girl to open her mouth. She complied, and his fear increased: instead of the clear pink of a healthy child's mouth, the back of her tongue and throat were gray. He scraped at them with a tongue

depressor, and the substance came off on the wood like a spreadable cheese. In medical school he had learned about the symptoms of diphtheria, but he had never seen a case until now. He remembered, with icy clarity, that it was highly contagious, particularly lethal to children, and impossible to treat without diphtheria antitoxin—which he didn't have. He barely even had any tongue depressors left. And the snow kept falling. If that toxin could not be delivered, thousands of people were going to die. With the snow blocking the every major route and the temperature dropping rapidly, Dr. Harberg had no idea how it could even be done.

*   *   *   *

The story is history now: how Nome was cut off by the harsh winter, before planes and helicopters and snowmobiles; how the fastest way to reach Nome was by dogsled; how Alaskans gathered diphtheria antitoxin in the town of Nenana and several teams of brave men and dogs worked in a relay to deliver the medicine. Thanks to their courage and endurance, Nome was able to save many of its children from diphtheria, and Balto, the lead sled dog of the first team to arrive in Nome, became an instant national celebrity. To this day in Alaska, dogsled teams run the Iditarod race from Nenana to Nome to commemorate this heroic feat.

Scientific Connection

Diphtheria is caused by Corynebacterium diphtheriae. These bacteria themselves, however, are only indirectly responsible for the symptoms of the disease. The actual damage is done by a protein they produce: diphtheria toxin, which destroys cells by preventing the creation of new proteins. The effect is similar to initiating a cellular labor strike: slipping diphtheria toxin into a cell just as the German government slipped Lenin back into Russia.

Proteins are assembled inside of ribosomes from amino acids that are gathered by tRNAs. The directions for how to make the protein are contained in the mRNA, which are like notes scribbled from DNA. Diphtheria toxin is extremely effective at disabling ribosomes, thereby preventing the production of new proteins. Without the ability to produce new proteins, the cell runs out of workers to maintain it. The result is cellular death.

Diphtheria toxin is so potent that a single molecule is capable of killing an entire cell. Corynebacterium diphtheria is inhaled and takes hold first in the back of the throat. Soon after these bacteria colonize, they begin to produce diphtheria toxin which kills the surrounding cells. The thick grey "pseudomembrane" in the back of the little girl's throat was the remains of all the cells that Diphtheria toxin had killed. If allowed to progress, the bacteria will spread into the blood and diphtheria toxin will destroy major organs like the heart and the kidneys; diphtheria kills 40% of all infected individuals who don't treat it.

Because diphtheria is communicated through inhalation, it's highly contagious and can quickly become an epidemic. Diphtheria was such a major health problem in the past that a vaccine was developed to protect against it. This vaccine introduces destroyed and inactivated diphtheria toxin molecules—known as diphtheria toxoids—into the bloodstream. A toxoid acts as a kind of "Wanted" poster for the immune system. Protein "bounty hunters" known as antibodies are created by the immune systems to seek out and destroy diphtheria toxin molecules. Antibodies remain in the bloodstream even after their target is eliminated and are continually produced, which is why vaccinations confer extended periods of immunity. If you were born in the U.S., you were almost definitely vaccinated against diphtheria when you were a small child. Antitoxins, on the other hand, are collections of antibodies made in an animal (like a horse) that are designed to track down and destroy a target protein; they give only temporary immunity, but are the treatment of choice when a protein toxin is already inside of your body. Due to the heroism of many men and animals the children in Nome were injected with the diphtheria antitoxin before it was too late.

Vaccinations and modern antibiotic therapy have largely eradicated diphtheria in the developed world, although it was a deadly and uncontrollable disease less than a century ago. However, there are many lessons to be learned from diphtheria toxin's mechanism of action. Bacterial ribosomes are different from those found in human cells and therefore represent tactical targets in the war against disease. Scientists have developed ways to turn this ribosomal inhibiting trick against the enemy with fantastic results. A huge number of modern antibiotics selectively target bacterial ribosomes in hopes of killing or at least containing them. This therapy has revolutionized the practice of medicine…

STORY 2: RED RASH OF THE ROCKIES

You are a pediatrician in Hoboken, New Jersey. You're surprised when one of your healthiest patients, an eight-year-old boy named Van, is brought into your office complaining of muscle aches and a terrible headache. He's had a fever and shaking chills for five days, and his mother tells you that she thought he had the flu until she saw his rash. It started, she says, on his wrists and ankles, spread to his palms and the soles of his feet, and over the past two days has crept up his legs and arms.

To narrow down the possibilities for these symptoms, you ask if Van has had any bug bites recently. Van pipes up, sounding more like his old cheerful self: "Yeah, me and Sarah were playing with grandma's dog and I got bit by three ticks and they stuck on me and they got all filled up with blood and we had to burn them to get them off!" Van's mom sighs and confirms this story, explaining that they were visiting family in the South Carolina countryside a couple weeks ago, where the dog and the kids ran around outside all day. As you suspected, Van has Rocky Mountain spotted fever; you prescribe the antibiotic doxycycline, which he must take for several days, and in two weeks his mother reports that he's in perfect health.

Scientific Connection

Van had a classic case of Rocky Mountain spotted fever. Despite its name, you don't need to be in the Rockies to catch it. You're most likely to contract it in the eastern and western coastal regions of the United States; the disease is most prevalent in the southeastern coast of the U.S. It is caused by a bacterium known as Rickettsia rickettssi, typically transmitted to humans through the bite of a dog tick. Bites can transmit disease-causing microorganisms to humans, so always mention tick bites or any other animal or insect bites to your doctor to help in your diagnosis.

Symptoms of Rocky Mountain spotted fever usually appear a week or two after the initial bite. Its early symptoms of fever, headache, malaise (feeling weak and tired), and muscle aches mimic the flu and are often missed. The telltale sign and the source of its name is the rash, which comes 2-4 days after the other symptoms begin. Rocky Mountain spotted fever is an extremely nasty disease that can quickly result in death due to failure of multiple organs. The spots represent tissue damage and though it is visible on the skin, these areas of damaged tissue are in many internal organs as well.

The disease would be fatal in many cases if it were not for modern antibiotic therapy. In a sense, doxycycline does to bacterial ribosomes what diphtheria toxin does to human ribosomes: targets them and shuts them down. In this case, ribosomal vulnerability works in humans' favor. Without a fresh supply of proteins to do the necessary order-promoting work, the bacteria are in big trouble: the inhibition is not enough to kill them, but they can no longer divide (reproduce) and are easily finished off by the immune system. Van's rapid recovery indicates that the bacteria are largely dead or at least are on the losing end of the battle, showing how effective this ribosomal inhibition is. To ensure that the bacteria are dead, patients taking antibiotics must always take the full course; if they stop when they begin feeling better, some bacteria may still remain and start reproducing again. These bacteria could be resistant to the antibiotic, which would make them much more difficult to treat. Drugs that disrupt bacterial protein synthesis are deadly weapons in any physician's arsenal and have led to the eradication and control of many previously fatal bacterial diseases.

⇝ *Take Home Message* ⇜

Ribosomes are essential to protein synthesis. Inhibition of ribosomes can have serious benefit in the case of antimicrobial medications or serious consequences in the case of toxins.

Cell Membrane: Shellfish Poisoning is for the Birds

THE STORY

Prince Edward Island, Canada, 1987.

"Hey Lewis, you want a mussel?"

Lewis looked at his friend's plate and made a face. "I don't know how you can eat those things. I'm good with my burger, thanks."

"Your loss. What's the point of living on an island if you don't take advantage of delicious fresh seafood?" Martin slurped a second mussel out of its shell and glanced up at the bar TV. "Oh, man, are you kidding me?"

"What's the matter now?"

"This is that friggin' movie where the birds all go crazy. Maggie!" he yelled to the bartender. "Can't we get the hockey game?"

"What are you talking about? This is a cinematic classic. Maggie, ignore this uncultured oaf. Watch for a minute, Martin, you'll see what I mean..."

...As the young blonde woman starts to unlock her car, something strikes her viciously in the face and knocks her off her feet. She touches her face and her hand comes away red with blood. What struck her was a seagull, now twitching on the sidewalk. Just as she starts to catch her breath, something shatters the passenger-side window: another bird has crashed through the glass. A full-grown pelican whizzes by her head, missing her by inches. It smashes at full speed into the wall of a nearby building and drops to the ground dead.

A shadow, then another, then many more cross her vision: looking up, she sees hundreds of birds in a disorganized, graceless swarm. Really frightened now, she manages to get the car un-locked and begins to drive home. Birds thump against the roof of her car: they seem to be dropping on the town like hail, slamming into pedestrians and buildings. The bodies of townsfolk litter the ground, cut and pecked in a dozen places as they lie dead amongst piles of bird corpses.

Then a bird smashes into her front passenger side window. She swerves off the road and crashes into a tree, and the screen goes black as her head hits the steering wheel. When she regains consciousness, her arms and legs are covered with cuts and her car is filled with the bodies of dead seabirds; panicked, she climbs from the car and flees toward her house, almost stumbling over more bird corpses. When she reaches the house, it brings no comfort: almost

every window is smashed, the front door is off its hinges, and the living room is filled with dead birds. She screams aloud ...

..."Lewis, you were right, this is a classic. Classically bad. Seriously, birds doing suicide dive bombs into people." Martin sucked down a final mussel and added the shell to his pile of empties. "Like that would ever happen. "Piranha" was better than this." ...

Scientific Connection

The plot of Alfred Hitchcock's *The Birds*—on which the movie Martin hates so much is loosely based—was itself based (also loosely) on actual events. In 1961 a large number of seabirds descended on Santa Cruz, biting people, crashing through windows, and smashing themselves into buildings. This aberrant avian behavior was discovered to be a mass case of domoic acid poisoning caused by toxic algae in the local waters. When the phytoplankton (plant algae) that produce domoic acid are prevalent in a marine area, they accumulate in shellfish like clams and mussels. These shellfish are a staple of many seabirds' diets; in Santa Cruz, eating shellfish tainted with domoic acid damaged their brains, causing them to lose control of their bodies and crash into objects and people.

Domoic acid acts on integral proteins that are embedded in the cell membrane of neurons (cells of the brain that transmit electrical signals). There are two major types of proteins that associate with the cell membrane: peripheral proteins and integral proteins. Integral proteins are embedded in the phospholipid bilayer and interact with both the extracellular (outside the cell) and cytoplasmic worlds while peripheral proteins only interact with one side. These integral proteins have many roles, one of which is to act as a gatekeeper: they decide which substances do and do not enter the cell. This is a very important job and failure to perform it can have dire consequences.

One of the most prominent gatekeepers of the nervous system is an integral protein known as the NMDA receptor. Its job is to regulate calcium ions ($Ca2+$): it lets a select number of them into the cell in order to promote the normal activities of the nervous system. The NMDA receptor has to keep a tight control on the amount of calcium it lets in, because the presence of too many calcium ions can cause the cell to destroy itself. These integral proteins are present in the cell membranes of many essential cells in the brain. When the NMDA receptor receives a signal from another cell, it opens its gate, lets a select amount of calcium in, and closes the gate. Domoic acid causes NMDA receptors to keep their gates open, allowing dangerous amounts of calcium to accumulate within cells and activating the self-destruct sequences of many cells in the brain ...

THE STORY

That was Tuesday night. On Wednesday, Martin wasn't at his desk. After giving him a few hours, Lewis called his phone, and his wife Jane picked up. "Oh, Lewis, I'm glad you called," she said, sounding distressed. "We're at the hospital. Martin came down with some kind of food poisoning last night. He had horrible stomach pains and he was really sick. He had some vomiting and diarrhea, but we didn't really get scared until he started having terrible headaches and shaking, you know, really trembling, he couldn't stop. So we went to the emergency room and it turns out it's crammed full of people with the same symptoms! The doctor thinks that is has something to do with shellfish poisoning—did he eat mussels with you last night?"

"He ate mussels, yeah. Listen, Jane, gimme a few minutes. I'm gonna take the afternoon off and I'll be right there. Tell Martin I'll be right there, okay?"

Martin smiled to see Lewis enter the hospital room. "See?" Lewis joked, relieved to see him sitting up. "I told you that you should have stuck with the hamburger."

"Hamburger? Oh, yeah, right, the hamburger." A look of distress came over his face. "You know, Lewis, it's the funniest thing. I remember the mussels, and I know why I'm in the hospital because Janie told me, but I don't remember feeling sick or driving here or anything that happened up till now. I can't remember the doctor that saw me, or any of the nurses. Janie keeps telling me the room number, but I keep forgetting anything that anybody tells me. I've never been here before and thank goodness you came because I don't think I could find my way out. The last thing I really remember is that goofy movie with the birds attacking. Man, I wish I could forget that."

Scientific Connection

Martin is suffering from amnesic shellfish poisoning. Domoic acid is known to cause serious damage to the areas of the brain associated with short-term memory, such as the hippocampus. Destruction of these areas causes "anterograde amnesia" which is the inability to remember new things after the poisoning occurred. This is why Martin can't remember anything that happened after he started to feel sick. The poison doesn't attack as severely the areas of the brain that retain long-term memory, so he remembers what his wife and best friend look like, and that he hates *The Birds*. But the brain damage that he suffered prevents him from forming new memories.

Domoic acid causes widespread damage to the nervous system as well, and can produce seizures (Martin's uncontrollable shaking). The poisoning can also damage the brain enough to cause coma and death. The toxin can take effect anywhere from 15 minutes to 48 hours after ingestion and usually begins with nausea, vomiting and diarrhea followed by signs of damage to the nervous system.

A well-known epidemic of amnesic shellfish poisoning occurred on Prince Edward Island in 1987 as a result of a toxic algal bloom. Over 107 people showed symptoms of having ingested the toxin, and a quarter of them had resulting short-term memory loss. Prince Edward

Island is world-famous for its mussels and the population boom of toxic domoic acid producing algae is definitely uncommon for this region. Marine wildlife are often indicators of toxic algal blooms; in 2002 in Los Angeles County, numerous marine animals showed considerable effects of domoic acid intoxication which indicated the trouble brewing in the waters. You can read more about that event here: http://whalerescue-team.org/rescue-stories/domoic-acid-poisoning-rescues/.

For more information on the mechanism of domoic acid intoxication: http://www.sciencedirect.com/science/article/pii/S0006899301032218

For more information on the 1987 Prince Edward Island mass in-toxication: Signs and symptoms of Domoic acid intoxication: http://www.ncbi.nlm.nih.gov/pubmed/1971709

⇀Take Home Message⇐

Integral proteins provide communication between the extracellular and intracellular environments and have important roles as receptors and gate keepers. If their normal activity is altered, negative physiological consequences can result.

Microtubules: Fungal Feeding Frenzy

FRAMING ANALOGY: INVITATION TO THE FEAST

Mayor Larry Hooper threw the morning paper off his desk in frustration. It landed in a heap on the floor, the headline still staring up at him: 'Fifth Shark Attack Victim in Brighton Bay; Officials Offer No Explanations.' At least this one had survived, he thought, albeit with one fewer limb than he had started out with.

Hooper turned his chair around. How many campaign promises had he made to grow the local economy? His efforts had led to a bustling boardwalk area by the beach, the piers now full of new restaurants and bars. But this bad publicity around the shark attacks was threatening to drive away all those necessary tourist dollars. The ocean had always contained sharks. Why did the sharks have to choose now to start eating his constituents?

"Beth, I'm going over to Doug's Fish Fry for lunch. If the press calls, feed them the same lines about ongoing investigations, using all our resources, blah blah blah."

"Will do, Mr. Hooper," Beth replied. "Enjoy your lunch."

Hooper chose his usual seat under a shaded trellis looking out onto the pier and the ocean beyond. He picked at his fried haddock, his mind still on the recent attacks. Some of the incidents had just been bites; gruesome and bloody, yes, but not too serious. Worse were the two deaths. A surfer had been bitten nearly in half, and they had to bury the torso of a local swimmer. His lower half was never found. Hooper shuddered just thinking about it.

It was a beautiful day, and the pier was lively. As Hooper scanned the activity, he noticed a family scraping their leftovers into the water before bussing their dishes back to the inside of the restaurant. He turned and saw a group of teenagers sitting at the end of the pier. A boy was playfully throwing French fries at one of the girls. Food scraps had accumulated in the water below. In the melee, a bowl of clam chowder was upended and joined the floating feast.

Hooper's brow started to furrow. His mouth opened slightly. That's when he heard the scream.

Just meters away from the group of teenagers, two men were pulling another man out of the water. His left calf was shredded. It looked like sloppy joe meat,

Hooper thought. A red cloud formed in the water below him. Hooper took out his phone.

"Beth, I figured it out. You need to get everyone in my office. Now. We can stop this."

"Slow down, Mr. Hoop-"

"It's the food in the water! Yeah, there have always been sharks in the water, but it's only since all the restaurants on the piers opened up that we started inviting them to dinner. Everyone's down here throwing their leftovers in the water, Beth. It's a buffet. We've actually been attracting hungry sharks to the Bay. It's no wonder some folks ended up on the menu, too.

"Write this down, Beth, because I've got the answer to our shark attack problem. First, we need to take care of the sharks that are already here. We've unwittingly trained them to come right up to the piers for food, and those sharks have got to go. Kill them, relocate them, whatever. Just get rid of the sharks that think my Bay is their own personal buffet.

"The second thing we have to do is stop feeding them! No more food waste in the ocean. Zero tolerance. I have the feeling that if we enact these two policies together, we might be able to save tourist season, maybe even save some lives."

Hooper looked on as EMTs approached the man with the mangled calf. He thought he made out a fin in the distant water, but it could have just been a cresting wave. He squinted his eyes at the ocean, and the predators who made it their home: "There's no such thing as a free lunch, you overgrown fishes. This restaurant is closed."

Scientific Connection

The primary problem in Brighton Bay is that people are being eaten by sharks. The reason that the sharks are present and causing trouble is that an environment has been created that has allowed them to propagate. People have been throwing food into the water, which has attracted a significant number of sharks into the area. The solution to the problem is fairly simple: the environment that attracts the sharks has to be eliminated and the sharks themselves need to be destroyed. If people stop throwing food into the water no new sharks will be attracted and if the present sharks are hunted then the waters will once again be safe. This situation is heavily analogous to the development of the fungal infection known as ringworm. Curing ringworm is a lot like curing the shark problem in Brighton Bay. It is essential to eliminate both the environment which favors fungal cell growth and any existing fungal cells that have taken residence. Elimination of the environment requires cleanliness but elimination of the fungal occupation involves a strategy targeted against the microtubules.

STORY 1: RINGWORM PINS WRESTLER

Aaron Gunderson is a 16 year-old wrestler on a winning streak. Unfortunately, Aaron is superstitious and believes that it's his singlet that's bringing him luck — so he hasn't changed it or washed it in four weeks! Aaron notices an itchy red rash developing on his chest, in his groin area and around his thighs. He would rather be itchy and revolting than lose a match. His doctor suspects ringworm, which is not a worm at all but a fungus, and she prescribes the antifungal drug griseofulvin. She also strongly recommends that Aaron wash his singlet and himself after every practice and match to prevent the fungus from growing out of control again. The drug, along with more sanitary clothing habits, gets rid of Aaron's ringworm. He feels so much more comfortable that his wrestling actually improves, and his winning streak continues. His multiple victories combined with his improved odor and lack of hideous skin lesions even lands him a date to the prom.

Scientific Connection

The fungal cells that cause ringworm live on the top layer of skin and consume constantly discarded skin cells. They like to live in warm, moist, dark areas, such as in between skin folds. Much like the inhabitants of Brighton Bay inadvertently created an appealing environment for sharks by discarding their food in the water, Aaron created a favorable environment for ringworm by continuing to wear the same tight, dirty singlet. The environment was very moist and skin cells that would normally flake off and get blown away were being stored in his singlet, which effectively became a fungal buffet. The fungi that cause ringworm are everywhere, but if you provide an environment that is full of things they enjoy eating (like delicious discarded skin cells) they will thrive and result in the symptoms of ringworm. Aaron's doctor suggested a two-pronged approach to treating his ringworm. First, Aaron needed to start keeping his singlet and his body clean to prevent the fungus from taking hold on his skin. Cleanliness would prevent the skin cells from accumulating and deny the fungal cells their free meal. Second, he was given the drug griseofulvin, which inhibits fungus growth by disrupting the formation and functioning of the fungus' microtubules.

Microtubules serve several functions in the cell, including maintaining cellular structure, acting as a conveyor belt for intracellular transport, and forming the spindle during mitosis. This last function is essential for cell division and it is griseofulvin's target. The fungal cells divide rapidly when they are in a favorable environment like a disgusting skin-cell-filled wrestling singlet. The division of the fungal cells allows them to overwhelm the body's natural defenses by generating more of themselves than the body can attack, resulting in the itchy inflamed rash of ringworm. Griseofulvin disrupts microtubules and prevents them from working. This results in an inhibition of mitosis, thereby preventing the fungus from reproducing and spreading. If the fungal cells can't divide, they can't overcome the body's natural clearing mechanisms and they are effectively eliminated. Cleanliness and proper hygiene eliminate the nutrition source by washing dead skin cells off while griseofulvin helps contain and kill the fungal cells that have taken up residence.

⤳ *Take Home Message* ⤙

Microtubules are essential for the process of mitosis. Drugs that disrupt microtubule function work by halting cell division in its tracks and are useful for eliminating fungal infections.

FRAMING ANALOGY: BATTERIES NOT INCLUDED

Christmas morning, 1987, 5:40 a.m.

Six-year-old Billy Walton squirmed in his bed and looked at the clock. His parents had told him the night before, firmly, that Santa doesn't come until 6 in the morning, and if he went downstairs to get his presents before then, the elves would take them back to the North Pole. But they just said he couldn't go downstairs. They didn't say he had to be asleep—and who could possibly sleep when the Terror Tank 8000 was waiting downstairs?

It was the toy of his dreams: a remote control tank that could climb walls, jump over holes, and fire both lasers and missiles at the same time. The first TV commercials for it came on the day after Halloween while Billy was eating his left-over candy and watching G.I. Joe. The day after that, Billy wrote the first of 54 letters to Santa—one for every day between Halloween and Christmas Eve—requesting a Terror Tank 8000. Some letters he placed in the mail; others he gave to any white-bearded, chubby old man he encountered, in case this was Santa in disguise, checking (as he always threatens to do) who's naughty and who's nice. He even stopped picking on his little brother, knowing that he and the Terror Tank 8000 could make up for it later.

Six o'clock! Billy leapt out of bed and catapulted downstairs. One box was bigger than any of the others under the tree; as his parents and little brother stumbled sleepily after him, Billy tore the wrapping off the glorious Terror Tank 8000, pried the box open and the toy from its wrappings, and pressed the button on the remote control.

Nothing happened.

While Billy burst into tears, his mother checked the label on the box to see what might have gone wrong. "Uh-oh," she said. "Batteries not included. We need

20 D batteries—18 for the truck and two for the remote control."

"It's not a truck, it's a tank!" Billy wailed.

"Hey, settle down," his dad said. "Settle down. Look at this, buddy, I came prepared." He produced two 10-packs of D batteries. "Help me get these in there."

"Mike, you didn't buy those at Hardware-R-Us, did you? I got an extension cord there the other day and it blew out my hairdryer." Together, Billy and his father installed the batteries and tried again to begin the Terror Tank 8000's first mission of mayhem. But the Terror Tank 8000 didn't climb the wall; instead, it bumped against it. It couldn't leap over the sink; it just fell in. It would go forward, but only slowly. The laser lights were dim and the missiles, instead of soaring across the room like in the commercials, just dropped. "Dad, why isn't it working?"

"I'm sorry to say that I think your mom is right—sorry, babe, that came out wrong. What I mean is that I bought these batteries for cheap. They look fine, but it seems they don't work so well. Just to be sure, let's check them in something else." He rummaged in the kitchen drawer and pulled out a flashlight, which worked fine with its present batteries; with the new batteries, though, its light was dim. "Well, buddy, there's your answer. The tank just can't get enough energy out of these cheapo batteries. The good news is that when stores are open again, we can get some regular batteries and the Terror Tank 8000 should be able to ride again."

Scientific Connection

Energy is the capacity to do work; the more energy you have the more work you can do. Of course, if you don't have any energy then you can't do any work. If there is an energy failure in a system, the observable sign is that elements of the system stop working or don't work as well as they should. In this case the work that Billy wanted to see was the fast movement, glowing lasers, and firing missiles of the tank. However, because the batteries didn't have enough energy stored in them, the tank could not do all of the cool things that it was supposed to. It moved, but slowly instead of fast; the laser lights were dull instead of bright; and the missiles didn't really work at all. In the same way, parts of a biological system working poorly or not at all can be a sign of an energy-source problem. The mitochondria are the battery of the cell: it is responsible for producing energy so that cellular work can be done. People with mitochondrial disorders have a set of bad batteries in pretty much every cell of their body. Usually tissues that require lower amounts of work to function tolerate this better, but tissues that require a lot of work to function (heart muscle, skeletal muscles, and nerves) suffer severely.

Mitochondrial disorders fit into a spectrum depending on how many diseased mitochondria the sufferer has per cell. This means that two people with the same disorder can have totally different disease processes based on the proportion of diseased mitochondria to healthy ones. Some people are barely affected while others die in childhood. Just as with the Terror Tank 8000, the proportion of good batteries to bad ones determines the overall functioning: more good batteries than bad means better functioning, and the opposite is also true.

One other fact makes mitochondria stand out: mitochondria are responsible for replicating themselves, so they have their own DNA. All of the mitochondria in your body originate in the oocyte, which is the genetic information you get from your mother's side. The proportion of good to bad mitochondria in that oocyte determines how severe the disease will be in your body. Typically mitochondrial diseases cause problems with multiple systems of the body; as noted above, the most striking abnormalities are associated with the muscles and nerves because those tissues require a lot of energy to function. Here are some examples.

MITOCHONDRIAL DISEASE 1

You are an ophthalmologist. A 20-year-old woman comes to your office and wants to know why she is losing her vision. In the past few months her vision has progressively decreased in both eyes and now she is nearly blind. Questions reveal that she's not the only one in her family whose vision has suffered: she has two brothers who both went blind much earlier in life. Her mother had progressive loss of vision, but never went completely blind.

Scientific Connection

This is a mitochondrial disease known as Leber's hereditary optic neuropathy (LHON). The target tissue is the nervous system, as evidenced by the degeneration of the optic nerve and the development of blindness. This disorder most frequently presents as a loss of vision in both eyes leading to blindness at a young age. This vignette also captures the spectrum of the disorder, with two brothers affected much earlier than their sister and a mother who has a slightly milder case. This means that the mother inherited fewer of the oocytes with diseased mitochondria from her mother; the early blindness of the two boys means that out of everyone in the family, they inherited the greatest percentage of diseased mitochondria per cell. The variability is due to the different numbers of diseased mitochondria in the oocytes that gave rise to each child. All the children are affected, as is the pattern of inheritance with mitochondrial disorders.

MITOCHONDRIAL DISEASE 2

A normally healthy nine-year-old girl is having an after-school snack at the kitchen table when she realizes that she can't move her right arm and leg. She wants to call for help, but she can't speak. She gets her family's attention by knocking a glass off the table with her left arm; they rush her to the hospital, where doctors determine that she has had a stroke. When tested, her serum lactate is 5.6 mM (normal is <1.5 mM) and questions about her family's medical history reveal that she had a little brother who died suddenly before his first birthday.

Scientific Connection

This is a case of MELAS syndrome: Mitochondrial myopathy, Encephalopathy, Lactic Acidosis, and Stroke-like episodes. Like LHON above, it is caused by mitochondria that do not work as well as they should. The mitochondria cannot make enough energy to satisfy the work needs of the nervous system.

This lack of energy caused the little girl's stroke. A stroke occurs when the cells of the brain don't have enough energy to keep working. The result is a loss of function, like the inability to move the limbs on half of her body and the loss of speech. She has a high blood level of lactate because the only way to make energy outside of mitochondrial mechanisms (oxidative phosphorylation) is glycolysis. In the absence of working mitochondria, lactate is the result of glycolysis, which is why the doctors tested for it in the case of this stroke. High levels of lactate can acidify the blood and disrupt the electrical workings of the nervous system and the heart. Her little brother also had this disorder, but he had more diseased mitochondria per cell and died of a lethal stroke at a very young age.

MITOCHONDRIAL DISEASE 3

You are a pediatrician. Your newest patient is a young boy whose guardian tells you that he has been periodically falling, twitching and losing consciousness. The boy is in the tenth percentile for height and shows signs of mental impairment. He has progressively lost his hearing and is now deaf, and your tests reveal that his vision has been deteriorating as well. The boy has difficulty walking and coordinating motions; his reflexes are perfect but his muscles are universally weak. You take a sample of his muscle tissue and examine it: under the microscope the mitochondria look clumped together and the muscle cells are ragged, as though they have been ripped apart.

Scientific Connection

This is a case of MERRF: Myoclonic Epilepsy with Ragged Red Fibers. It is a mitochondrial disease that typically presents with seizures, ataxia (inability to coordinate muscle movement), deafness, vision loss, muscle weakness, stunted growth and poor mental development. The tissues with high-energy demand, like the nervous system and the muscles, are heavily affected. The loss of hearing, loss of vision, seizures, poor mental development, and loss of coordination are all evidence of energy failure in the nervous system. The weakness of the muscles and their torn appearance under the microscope is evidence of energy fail

-ure in muscle tissue. The muscle cells don't have enough energy to do their job and survive so they choose to do their job and die, resulting in their ragged, ripped appearance.

MITOCHONDRIAL DISEASE 4

A woman brings her young son in to your office because of frequent and recurring infections. The boy is short for his age. His mother says that he is winded every time he goes upstairs and that he tires easily in general. Your questions reveal that she had a brother and an uncle who had similar symptoms. A blood test shows extremely low numbers of white blood cells, which explain his frequent infections.

Scientific Connection

Barth syndrome is another mitochondrial disease. It's characterized by increased risk of infection, rapid fatigue and constant feeling of being tired, delayed growth, learning disabilities, and poorly working heart. Death usually is due to heart failure or infection. Once again, this disease shows a pattern dominated by energy failure in muscle tissue. As a result there is generalized muscle tissue weakness, evident in both the skeletal muscles and in the heart. Unlike the previously discussed mitochondrial disorders, the inheritance pattern associated with this is X-linked recessive, which is why the mother's uncle and brother are affected as opposed to her and all her children. While mitochondria are responsible for replicating themselves and making most of their own proteins, some of their proteins are made from information in the nucleus, which accounts for this pattern of inheritance.

MITOCHONDRIAL DISEASE 5

You are a first-year medical student doing a physical on a woman in a hospital. When you look into her eyes you notice an unusual color in both of her retinas. You ask her to follow your finger with her eyes and she can't move her eyes from side to side. When you ask her to stand up and walk to the other side of the room, she does so with a wide unsteady gait like she is drunk. Her muscle strength is poor and she has a history of heart problems.

Scientific Connection

This woman has Kearns-Sayre Syndrome. It is characterized by abnormal retinal pigment accumulation, opthalmoplegia (inability to move eyes from side to side), myopathy (generalized muscle weakness), ataxia (poor coordination), heart problems, and diabetes. This disease has all the hallmarks of mitochondrial dysfunction including reduced function of the nervous system, skeletal muscles and heart.

➥Take Home Message⬅

Mitochondrial diseases cause energy failures in multiple tissues. The tissues with high energy demands suffer the most dysfunction. The skeletal muscle, cardiac muscle, and nervous tissue are the most frequently affected. The degree of severity depends on the proportion of diseased to healthy mitochondria per cell.

FRAMING ANALOGY: LIVING IN SQUALOR

"That's it. We're turning around."

Karen Kennison slammed on the brakes and the car began to spin out. "Pump the brake, Karen!" Karen did, and the car slowed and stopped halfway across the road. Fortunately there were no other cars to be seen. There hadn't been for miles. The snow continued to fall. "What was that supposed to be?" Karen's new fiancé Gabriel demanded. "You've been acting weird ever since we started driving. What is it? You don't want me to meet your folks? We've been dating for two years, I thought maybe it was time, and they never come down to see us."

"I told you they don't leave the house much."

"Yeah, I know, and you said they save everything, but it's really no big deal, Karen. I've seen messy houses before. I mean, I live in one, or I did before you started making me pick up after myself." He kissed her cheek to show that he didn't mind it. "Come on, let's get back on the road."

"It's because of them that I'm like that. I tried to explain before we came up here, but I can tell you won't believe me till you see it, so let's just go. We're pretty close now, it's just up this big hill."

Twenty minutes or so later, they pulled into the driveway of a one-story house that looked pretty ordinary except for the groceries laid out on the front stoop. Karen took a deep breath. "Okay. I'm sorry. I warned you," she said, and opened the door.

The smell of unwashed humans and rancid milk came out at them in a wave. Gabriel gagged, but Karen seemed unaffected and led him into the house. Led him between stacks of newspaper, paper bags filled with garbage, paper bags filled with old towels, furniture in a lean-to of pizza boxes. Karen's mother stepped out from behind a stack of full black garbage bags. "Hi, you must be Gabriel! I'm Martha. It's so nice to finally meet you! Sorry about the smell—we misplaced some milk a couple of weeks ago and we haven't found it yet. The good news is that the worse the smell gets, the easier it'll be to find."

"Ah—is this your restroom?" Gabriel said, just as Karen reached to stop him, shouting, "No, Gabe, don't open—" An avalanche of wrappers, papers, old socks and heavy boxes, smelling like mildew

and old sweat, crashed down over him. "There's something moving around in there," Gabriel sputtered when he got to his feet, "I felt it moving, some kind of animal. Do you guys have a cat? It might be hurt."

Martha shook her head. "We don't have any pets, unless you count the rats and squirrels. Dan, is that you at the door? Did the pizza man come? Come on in here and meet Karen's fiancé."

"Dan Kennison," said Karen's father from behind the pizza box. "Nice to meet you, Gabriel. We don't do a ton of cooking around here, mostly because we can't use the kitchen. We need the fridge and the oven and so on for our files."

"There's a pizza place that delivers," Martha put in, "but they won't drive up the hill, so we meet them at the bottom of the road. See if you can find some plates, Gabriel." Shuddering, Gabriel went into the kitchen, only to see the counters full of greasy wrappers, old bottles, used sponges and more stacks of paper. In the oven—he peeked in to see—was a box labeled "tax returns."

He returned to the front room—he refused to call it a living room—and said, "I'm sorry, I couldn't find any plates. I saw you had some groceries outdoors—so you use the outside as a natural fridge in the winter?"

"That's right! Good detective work, son," Dan confirmed. "It's a pretty good system unless snow buries the food. Then it's a treasure hunt. Of course sometimes animals get into it, but they need to eat too, right? Have some pizza." Gabriel took a slice, wincing, trying not to think about what he'd touched when the trash fell on him. "You've had a long drive," Martha was saying, "so you just let us know when you're ready for bed and we'll move some stuff off the fold-out couch. It's plenty big enough for four."

"What about ... bedrooms?" Gabriel couldn't help asking. Martha blinked. "Oh, we use the bedrooms for storage, of course," she said, just as all the lights in the house went out.

"Dang!" said Dan. "Someone'll have to go down and flip the breaker."

"I'll get it, Dad," said Karen, standing up. Dan looked reproachfully at Gabriel. "Now, son," he said, "are you going to make that poor little girl go down to the basement by herself?"

Gabriel thought of a couple of things to say—that it was Karen's family's house and she knew where the breaker was; that it was the Kennisons' house, and they knew even better where the breaker was; that Karen was 5 foot 8 and trained police attack dogs for a living; that he would rather walk away into the snowy wilderness and freeze to death than see what the basement looked like in that house. Instead, he got to his feet and asked, "How do I get into the basement?"

"Well, that's the tricky part. We store our old bikes and fans and cereal boxes in the basement stairwell, so what you're going to have to do, Gabriel, is go outside around the back of the house. There's a window there that opens into the basement, so you can open that window—or break it, doesn't matter much to us—and climb on down. The breaker should be a few feet to the left of that window. Take this flashlight—it gives good light and if anything's living down there, just give it a whack." Trembling with disgust and horror, Gabriel pushed the front door open—gasping with relief at the smell of the fresh air—and trudged out into the snow.

Scientific Connection

Possibly because of some form of mental illness, the Kennison family is living in squalor and it is seriously affecting the quality of their life. They continuously create garbage in their house and never remove it. As a result the trash accumulates in their home, making it virtually uninhabitable. The basic functions of the house have been totally disrupted by the massive amount of refuse within it: they can't cook in the trash-filled kitchen, so they eat takeout (generating more trash) and keep their food in the snow. They can't sleep in the trash-filled bedrooms, so they and their guests have to sleep on a fold-out couch surrounded by trash that may or may not be infested with rats or squirrels. They can't get down the trash-filled stairs, so the only way to solve an ordinary problem—an internal power outage—requires crawling through small places and fighting wild animals with a flashlight. They can't use the trash-filled restroom, and you can use your imagination to guess what kind of "solution" they might have found for that problem. The main idea is that when a house becomes full of junk it is miserable to live in. If you can understand this basic concept, you can understand the basis of any lysosomal storage disease.

In many ways a cell is similar to a house. When you have garbage in your home, you take it out to a collection site like a dumpster or a trash can. Cells also accumulate waste products that need to be disposed of, and the lysosome is a cellular organelle that functions as a "garbage collector." The proteins that work within the lysosome have the ability to break down and destroy almost any intracellular material. When waste accumulates inside of the cell it gets picked up by a lysosome; proteins within the lysosome pounce on the refuse and tear it into little pieces, which are then recycled and made into new cellular components. If any of the proteins that work within the lysosome (lysosomal enzymes) are bad at their job, the intracellular trash can't be degraded. It will accumulate within the cell, causing the same degree of disorder that you see in the Kennison household.

It is just as bad to have cells full of trash as it is to live in a house full of trash. Lysosomal storage diseases can be fatal and cause serious problems in the tissues they affect. The junk-filled cells often increase in size, leading to an enlargement of whichever tissue they happen to be part of. This results in organ dysfunction, enlargement, and considerable deformity in many cases.

⤙ *Take Home Message* ⤚
Lysosomes are responsible for destroying waste within a cell. When lysosomal enzymes don't work or aren't present, waste accumulates in cells, leading to possible tissue enlargement, poor cellular function and disease.

STORY 1: POMPE'S DISEASE

The year is 1980. A young mother and father bring their baby to his two-month checkup with a great deal of concern. The infant shows considerable signs of weakness. The mother describes him as "floppy" and "like a wet washcloth", and says that he barely seems to have the strength to breastfeed. The baby demonstrates relatively low levels of activity, is generally sluggish, and frequently lies still. He is also very small and light compared to other babies of the same age. When the doctor feels the baby's abdomen his liver is much larger than normal. Imaging tests show that he also has a massively enlarged heart. Unfortunately, the doctor has to give these parents bad news: nothing can be done except treatments to support the baby's failing organs. He dies of heart failure before reaching his first birthday.

Scientific Connection

The baby had Pompe's disease, which comes from an inability to break down glycogen. Glycogen is a storage form of glucose, like a candy bar that you keep in your back pocket in case you get hungry. If you continue to make glycogen and never break it down then it will accumulate within the cell like trash in the Kennison's house. Glycogen is stored in both muscle tissue and the liver; the baby's enlarged heart and liver were the signs of glycogen accumulation. The muscle cells of the heart became so full of glycogen that the tissue began to fail and the infant died of heart failure. The muscle cells of the body were probably filled with glycogen and failing as well, since the baby was weak and found it difficult to move or eat.

Pompe's disease results from poorly functioning lysosomes. A protein called α-glucosidase works within lysosomes to break down glycogen. In most people α-glucosidase is an excellent lysosomal employee: when glycogen enters the lysosome it is promptly broken down and never ends up accumulating. However, in people with Pompe's disease α-glucosidase is defective and can't do its job; as a result, glycogen accumulates within cells producing symptoms of the disease. Not all glycogen is broken down within lysosomes, but they play a vital role, as shown by the damage that Pompe's disease can cause.

Pompe's disease represents a spectrum of disorders based on the degree of incompetence α-glucosidase exhibits. The most severe form of Pompe's disease is the infantile form depicted above, which kills within a year and a half after symptoms begin. In that case α-glucosidase can't do anything at all. In the past, severe cases of Pompe's disease were rapidly fatal and there was no treatment, but modern medicine has the potential to change the course of this once horrific affliction.

STORY 2: CLEANING HOUSE

The year is 2018. A newborn screening test determines that a one-week-old baby girl has Pompe's disease. Administration of the drug Myozyme begins immediately. Two months later, the baby's parents bring her back to the hospital for a follow up visit. She seemed to be happy and active with good muscle strength. Her mother says that she's feeding well, and her weight and size are normal for her age. When the doctor feels her abdomen, the liver is of normal size and the imaging studies show a normal-sized heart. The doctor is happy with the baby's progress and says that she will have to remain on Myozyme for the rest of her life to prevent her from developing symptoms of the disease.

Scientific Connection

If you are like the Kennison family and cannot manage to clean your house, you may be able to hire someone to clean it for you. The baby girl in this vignette has the same problem as the baby boy in the first vignette. Both babies have a problem with their lysosomes: the deficiency associated with α-glucosidase. Why is she healthy while he is dead? The answer is that the little girl has "hired" a good version of α-glucosidase to clean the glycogen out of her cells. The drug Myozyme is actually a recombinant form of the protein α-glucosidase.

The newborn screen is a test done to determine if an infant has any of a list of serious diseases that need to be caught and controlled early. In the future Pompe's disease might be included on that list, but it isn't now (hence the 2018 date on this story). In this case, the newborn screen determined that this little girl had Pompe's disease before she was one week old. The physicians knew that her α-glucosidases were incompetent and had no chance of degrading glycogen. That in turn meant that glycogen would accumulate in her cells and she would be unlikely to live long, so they prescribed the drug Myozyme: this artificial α-glucosidase is able to enter the lysosomes and degrade the glycogen. As a result the little girl never developed the signs of the disease and remained healthy while the untreated boy died.

The little girl will always produce incompetent α-glucosidases. If she doesn't take Myozyme for the rest of her life, her cells will become full of glycogen and she will develop the disease. Myozyme is a current drug that has made a huge difference in the treatment of Pompe's disease and its development was chronicled in the motion picture Extraordinary Measures. If you want to know more about Myozyme and Pompe's Disease look at this document: https://www.lsdregistry.net/pomperegistry/hcp/safety/preg_hc_s_productinfo_smpc.pdf

➥*Take Home Message*⬳

Lysosomes are responsible for degrading molecules inside of the cell. If one of the proteins that works within the lysosome is defective then the indigestible material will accumulate in the cell leading to disease.

Peroxisomes: Radical Behavior

FRAMING ANALOGY: RADICAL INITIATION

"How's it coming in there, Gabe?" Karen called into the kitchen, smoothing the tablecloth down over the big but somewhat wobbly dinner table.

"The roast has about an hour to go, the potatoes are in and the rhubarb wine is chilling," Gabriel shouted back. "I just have to make a salad. Listen, Karen, I really appreciate you having my brother's family here for Christmas. I just hope you don't change your mind about me after you meet them. Especially Billy ... oh Lord... I almost managed to forget he was coming. There aren't enough exorcists in the world to deal with that little devil."

"Please, sweetie, it's the least I can do," Karen cooed. "You were so good with my parents at Thanksgiving. You kept your cool the entire time, even when you had to get those rabies shots right in your stomach. I owe you holiday parties for the next fifty years."

"Well, hopefully for the next fifty years we'll be having them together." Gabriel would have pulled her in for a kiss, but the doorbell rang. Karen went down to the door and let in an older, plumper version of Gabriel, a pretty but tired-looking woman, and a little boy with an innocent face. "You must be Billy," Karen smiled at him. "What's that big toy you have there? It looks really heavy."

"This is the Terror Tank 8000!" Billy informed her. "Santa brought it! It's the best toy ever! It goes really fast and it destroys everything in its path! With lasers!"

"That's ... nice. Come on in, Mike, Alicia, let me take your coats. Billy, your Uncle Gabriel told me you like chocolate-covered raisins, so I got some just for you—if your mom and dad say it's okay to have them before dinner?" Billy's mother Alicia mouthed over Billy's head, "Anything to keep him quiet," and said more loudly, "You can have a few, Billy, and say thank you." Billy dove at the bowl and shoved a handful of chocolates into his mouth. He kept eating them while the grown-ups lent a hand in the kitchen, and then announced, "Those were weird raisins. Kinda crunchy. But really good. Can I have some more?"

"Why don't you take it easy, buddy?" his father suggested. "Save some room for dinner."

"Oh, Mike, it's Christmas," Karen laughed, "and he's a guest. Let him eat as much as he wants. Billy, I'll get you some more. Alicia, can I get you anything to drink while we wait for dinner to be ready?"

An hour passed, and constantly refreshing Billy's supply of chocolates kept him surprisingly quiet. The roast came out of the oven, and the adults began carrying food from kitchen to table. Just before they were about to sit down, Gabriel darted back into the kitchen. Karen followed him. "Gabriel, everything is perfect. Seriously. Come sit down."

"I just wanted to get dessert out so we wouldn't have to deal with it later. Have you seen those chocolate-covered espresso beans your dad sent?"

"They're right here, in this bowl." Karen pointed. "Come on, aren't you hungry?"

"Those don't look like them. I think those are Raisinets."

"Raisinets?" Karen stared. At that moment, they heard the revving of an electric motor, followed quickly by the sound of glass breaking, a heavy thump, and a child's voice shrieking, "Nothing can stand in its way! All bad guys run for cover!"

Karen and Gabriel joined Billy's parents in their run for the bedroom, where Billy's voice had come from. The Terror Tank 8000 flew out to meet them, crashing into Gabriel's shin. He sank to the ground and moaned. Billy had blocked most of the bedroom doorway with one of the chairs, and the sound of broken glass had come from the full-length mirror. He had given himself camouflage makeup with Karen's eyeliner. He shouted, "You'll never take me alive!", waved the remote control, and sent the Terror Tank 8000 down the hall straight toward the dinner table. It crashed into the table leg. The table wobbled and went down. The roast, the potatoes, the rhubarb wine, the peas, the stuffing, the gravy, the dishes, the glasses, the silverware slid to the floor.

Gabriel half-stood, still clutching his shin, and looked at his brother.

Mike sighed. "I think we passed a pizza place on the way here that was advertising Christmas night delivery."

"You go call," Alicia said. "I'll help with the cleanup while we wait for King Kobra Kommando here to wear himself out. You want to give him some space when he gets like this. He has a tendency to bite and his fingernails are a lot sharper than they look. Karen—we're so sorry—where do you keep the mop?"

"My apartment," Karen murmured, helping Gabriel the rest of the way up.

"Your beautiful dinner."

"You still want to marry me?" Gabriel managed to ask through pain-clenched teeth. "I think that little monster broke my tibia."

"More than ever," Karen assured him, kissing him. "But between your family and mine, it'll be some wedding. We're going to need a lot of rhubarb wine."

Scientific Connection

Reactive oxygen species are the molecular equivalent of little Billy. They represent a phenomenally dangerous consequence of living in an oxygen-rich environment and are the bane of many biological systems. Reactive oxygen species, also known as oxygen free radicals, contain oxygen atoms with a single unpaired electron. Radicals will go to any length to get another electron to correct their imbalance. They will steal electrons from the nearest molecule, and as a result they are highly reactive—and the reactions they provoke are violent on a cellular scale. Reactive oxygen species have the capacity for phenomenal amounts of destruction: they can tear giant holes in a cell's membrane, cripple organelles, interfere with proteins trying to do important jobs or even rip the whole cell apart, just like Billy trashed the apartment in an unstoppable caffeine-induced frenzy, even though he was much smaller than the apartment.

Reactive oxygen species are even used as weapons by the immune system. Bacteria are captured and caged inside white blood cells before being dunked in a bath of radicals in the hopes that it will kill them. As long as the body performs oxidative reactions and we live in an oxygen-rich environment, reactive oxygen species will remain a persistent biological problem. Fortunately we have developed many defenses against these chemical berserkers and one of them is the peroxisome.

Within peroxisomes, the protein catalase works to prevent free radicals from forming by performing this reaction: $2H_2O_2$ (aqueous)$\rightarrow 2H_2O$ (aqueous) $+ O_2$ (gas). While Hydrogen peroxide ($H2O2$) is not in itself a radical, it can become a radical very easily and cause massive damage. In this story Billy was like hydrogen peroxide in that when he entered the house he was (relatively) harmless; however, after ingesting hundreds of caffeine-loaded chocolate-covered espresso beans he went radical and began his reign of terror.

⇢ *Take Home Message* ⇠

Reactive oxygen species have a phenomenal capacity for destruction. Peroxisomes are involved in the neutralization of reactive oxygen species through the activity of catalase.

THE STORY: HYDROGEN PEROXIDE AND WOUND CARE

One minute Clifton Brown, age 9, was coasting down the hill on his new board and thinking about his next jump. The next, he was lying on his face on the pavement with his knees in a mud puddle. He forgot about being the next Tony Hawk and limped back up the hill to his house.

"Ooh, that looks nasty," his mother said when he came into the kitchen where she was doing dishes and watching her Saturday soap operas. "What did I tell you about that skateboard? Now come here." She grabbed a clean dishcloth, squirted soap on it and rubbed it across his knees. "Ma," he wailed. "Okay, that's enough!"

"No it isn't. I still see some dirt in there." Clifton watched in horror as she got the dreaded bottle of drugstore hydrogen peroxide down from the top shelf and poured it over his scrapes and cuts. It stung and bubbled, but when his mother wiped his knees again, the dirt came away, displaced by the bubbles. "Let's cover that up now," his mother said, pasting a Band-Aid on, "and then you can go back out, but if you fall off that board again, don't come crying to me."

Scientific Connection

Those bubbles are visible evidence of catalase hard at work. When hydrogen peroxide is poured on the wound it passes into the peroxisomes and is defused by the catalase in this reaction: $2H2O2 \rightarrow 2H2O + O2$. The bubbles are the $O2$ gas produced by the reaction. This bubbling action is a great way to clean out the deep crevices of a wound, but for killing bacteria, hydrogen peroxide is not the best choice because aerobic bacteria—those that live in oxygenated (oxygen-containing) en vironments—often have catalases of their own. As a result, they can deactivate hydrogen peroxide before it can become the reactive oxygen species that would kill them. Anaerobic bacteria, on the other hand, fall prey to hydrogen peroxide quite easily because they live in environments without oxygen and do not tolerate reactive oxygen species well.

Iodine and rubbing alcohol are much more effective at killing bacteria than hydrogen peroxide is. The prevailing thought is that soap and water are better for cleaning out a wound than hydrogen peroxide, which (like rubbing alcohol) can damage tissue and slow some aspects of wound healing. However, it is much worse to have a dirty wound than a little bit of tissue damage. Dirty wounds are very dangerous because dirt is loaded with bacteria, especially anaerobic bacteria, which can cause infections and deadly diseases like tetanus. Infections can easily kill a normally healthy person if they get out of control. In Clifton's case, the combination of soap and water (to destroy some bacteria) and hydrogen peroxide (to destroy anaerobic bacteria and push the dirt out of the wound) was probably the safest way to go.

⇀ Take Home Message ↽
Wounds need to be cleaned to facilitate rapid healing and to prevent infection. Hydrogen peroxide is good for cleaning wounds and killing anaerobic bacteria through its transformation into reactive oxygen species, but washing dirty wounds with soap and water is the recommended way of cleaning them.

THE STORY: A MATTER OF CONCENTRATION

Craig and Dwayne went straight from morning football practice to first-period chemistry. "Hey bro," Craig said as they were setting up the lab on solution and suspension, "we should put some of this on our arms." He held up a stock solution of hydrogen peroxide. "This keeps stuff from getting infected, right?"

"Yeah, my Nana used to put it on me when I would scrape my knees. Good call, bro, but wait till Mr. Potato Head isn't looking." Mr. Potato Head was Mr. Potter, the chemistry teacher. When Mr. Potter was helping another pair with their experiment, Craig held out his arm and whispered, "Lay it on me, bro. Ow, that stings."

"I think it's supposed to," Dwayne started to say, but stopped in horror as the skin around the cuts began to turn white. "I don't think it's supposed to hurt this much," Craig whimpered. "Ugh, it itches, it burns!" He was really yelling now, and kept yelling while Mr. Potter dragged him (by his other arm) to the safety shower.

"What do you think you're doing, messing with chemicals like that?" the teacher demanded. "Keep your arm in the water!"

"I just didn't want our cuts to get infected," Dwayne whined. "I thought you put hydrogen peroxide on cuts."

"A three percent solution! Not a fifty percent solution! Obviously you goofs haven't been paying attention in class!" Although the burning had stopped, Craig continued to whimper as blisters rose up on his arm where the hydrogen peroxide had touched. "Both of you have detention with me and just in case you haven't learned your lesson, we're going to go over solutions and percentages until you get it right!"

Scientific Connection

A chemistry lab has the potential to be a dangerous place. Chemical reagents should be treated with respect and used according to instructions—don't improvise! Many people have been seriously hurt or killed by the misuse of laboratory equipment and reagents. The hydrogen peroxide solution purchased in stores is typically a 3% solution, which is already concentrated enough to damage tissues through the generation of reactive oxygen species. A 50% solution has significantly more destructive capacity and would generate enough reactive oxygen species to tear apart anything in its path. If this concentration of hydrogen peroxide were poured on your arm it would ravage the tissue, resulting in burning and blisters. Hydrogen peroxide leads to the generation of oxygen free radicals and those are what cause the massive tissue damage. A 50% hydrogen peroxide solution would overwhelm any defenses that you have against free radicals. The key to treatment is getting the substance off as fast as possible, so if you're accidentally exposed to a caustic (burning) reagent, you should hold the exposed surface under running water for several minutes. "The solution to pollution is dilution."

⇒ *Take Home Message* ⇐

Reactive oxygen species are extremely destructive. Chemistry labs are potentially dangerous places and should be respected.

Peroxisomes: A World Without Peroxisomes

THE STORY

A newborn baby has been having frequent episodes of uncontrollable shaking. The baby is extremely weak and does not move around with any excitement or vigor. Its weakness prevents it from being able to effectively breastfeed. The infant does not seem to respond to any visual or auditory stimuli, which suggests that it's both blind and deaf. The infant has a number of unusual facial features including a high forehead, a very small jaw, abnormal looking eyes, and a very broad nose. Tragically, the infant dies within the next few months.

Scientific Connection

This child died of Zellweger syndrome, a serious disorder that is lethal within the first few months of life. As you might guess from the description of this case, this child has numerous problems with multiple systems of the body. The infant is having seizures, it can't see or hear, and it is extremely physically weak. While the sequence of events that lead to these numerous abnormalities and characteristic facial features is currently unknown, the cause of it all is defective peroxisomes. Peroxisomes are organelles that have multiple functions within a cell. These include preventing damage caused by oxygen radicals, as well as playing a critical role in converting certain fats into biological energy. The peroxisome itself is like a little factory where both of these processes take place. A factory, however, is useless without workers inside to do the labor. A steel mill with no steelworkers produces no steel, and a shirt factory with no stitchers or cutters produces no shirts. The physical effects of Zellweger syndome are caused by peroxisomes with no proteins to work inside of them them. Zellweger syndrome is the most severe of a class of disorders known as peroxisomal biogenesis disorders: the cells contain peroxisomes, but the proteins necessary to staff them never make it to work. Instead, these proteins can be found wandering aimlessly around the cytoplasm. Zellweger syndrome is a consequence of having no or very few functioning peroxisomes and its severity is a sign of just how important peroxisomes are to the proper workings of the body.

⇀ Take Home Message ⇀

Peroxisomes perform important functions related to the minimization of oxidative damage and the utilization of certain fats to make energy. The physiological effects of non-working peroxisomes are extremely severe and often fatal.

Rough Endoplasmic Reticulum: Riki Tiki Tavi

PART 1: AMY

"Well," Meena said cheerfully, "this is it. Look at the nice garden!"

"Whatever," Amy mumbled, staring through the gate. "Is it locked?"

"That was the housekeeper I was talking to on the phone, when we were in the cab. She should be here any second to let us in." As she spoke, a short, stout woman in salwar kameez came to the gate and unlocked it. "Welcome, Mrs. Kipling!" she said. "I'm Barathi. And this must be your little girl!"

Normally Amy would have sworn at anyone who called her a little girl, but she had something more important to think about. "Pleased to meet you," she said politely in Hindi, and then in English, "Do you think I could have some milk or something?" Barathi waved them into the house, still talking, but Meena interrupted her. "Barathi, that driver seemed to think this house was bad luck. He said that four people have died here recently."

Barathi looked uneasy. "These drivers are all ignorant people, Mrs. Kipling. He probably just didn't want to drive you all the way out here. Mr. Krishnan wouldn't have sent me to work here if there was any problem like that." Amy was only half-listening; as soon as she got her glass of milk, she dragged her suitcase into the small bedroom, set the milk on the floor and opened her carry-on bag.

A nose and whiskers emerged, followed by a long thin body. "Here you go," she whispered. "Are you hungry?" The animal sniffed the milk and began to lap it up, dipping its long pointed nose.

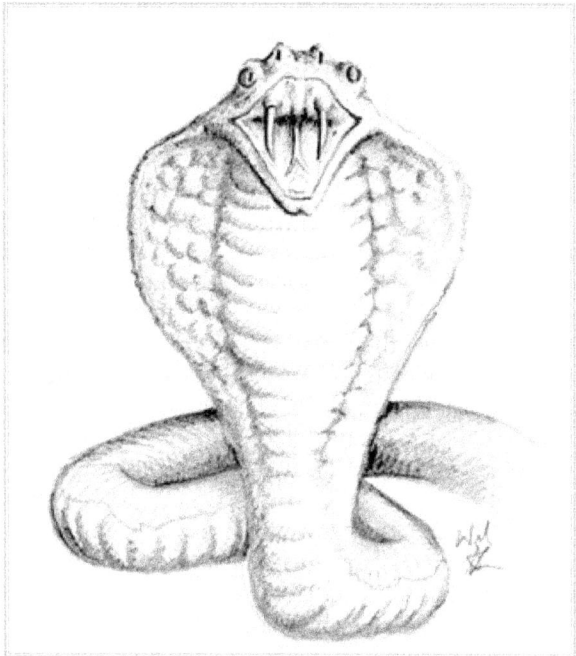

Amy chattered as it ate. "What were you doing in that puddle, you poor thing? You look just like my ferret Marilyn. I had to leave her at home because India's stupid government won't let people bring animals in from other countries. Mom said. I don't know why she had to take this stupid job anyway. Whenever I even say anything about it she just asks me if I want to pay for college myself. Or why couldn't I stay in St. Louis with Dad? Mom says she wants me to see where she grew up, but there's not even anything to do here.

I'm glad I found you. You can keep me company, but I have to keep you hidden or Mom will freak out, okay?"

When the animal had drunk as much of the milk as it could reach, it sat up, cleaned its whiskers, and made a chattering, clicking noise. "Rikki!" said Amy delightedly. "Are you telling me your name? Is your name Rikki?" The animal nosed its way under her hand.

PART 2: RIKKI

The mongoose was delighted to have a human again. Raised by a man who kept him around to eat bugs and the occasional snake, he had come back to the shack one day to find it empty. There is plenty of garbage in Mumbai, but human scavengers are bigger and sometimes faster than a mongoose, and he had been close to starvation when Amy found him. She pet him, brushed him and brought him table scraps; he listened to her human words which, of course, he couldn't understand, but her tone was affectionate. It was a good life.

A month after coming to live with Amy, he was taking a twilight stroll around the edge of the property fence when he smelled a rat—actually, two. In his old home rats had been the enemy, so he took an aggressive stance and called, "Who goes there?"

"Sorry, man, we didn't know there was anyone here." The tiny footfalls came closer. "We're the Gray Sisters. We're not looking for a fight. But what are you doing in the humans' garden? Are you crazy?"

Rikki didn't want to admit that he was a pet, so he said, "I'm not afraid of humans."

The larger of the two Gray Sisters snickered. "Nobody's afraid of humans! Don't you have a nose? Haven't you smelled Naga and Nagini?"

"Who?"

"City boy!" sneered the smaller rat. "Naga and Nagini are cobras. They're about as big around as you are, and they own this place. They're nesting under the back porch and they're not afraid of any humans. They've already taken out four and they're probably itching to get a fang on the ones that just moved in. They don't want anyone messing with those eggs!"

"Did you try?" Rikki inquired, knowing that rats love to eat eggs—almost as much as snakes love to eat rats. "Or were you too scared?"

The larger rat hissed, "Don't you joke about that, Skinny, we lost our sister just last week. Take my advice and get out while the getting's good. Forget about these humans—they're as good as dead." The two of them scuttled away through the fence. Rikki stood very still in the grass, thinking hard. Killing snakes is what a mongoose does best, but he'd never tackled one that big. But maybe the Gray Sisters had just been trying to scare the newcomer? Rikki crept toward the back porch, keeping downwind so the cobras wouldn't smell him, and peeked underneath.

At first he only saw darkness; then gleams of light from the house began to pick up their coils, and he smelled the unmistakable musty smell of snake. "We will need to bite them several times," he heard a cold, dry voice say, "and gently, or they will wake before the venom stops their breathing."

"We should bite the large ones first, then," hissed a second voice. "If one of them wakes, it could be strong enough to kill us, and then who would protect the eggs?"

"Humans are slow and weak. I have no fear for myself. We will bite the small one first, that will be easiest and leave plenty of venom for the others." Rikki froze in horror as he realized who the cobra meant—then leaped back as first one, then the other began to uncoil and head right for the hole he was watching through! Scampering to a safe distance, he watched, sick with fear and rage, as their long bodies emerged and slipped toward the front of the house. They were almost as big around as Amy's arm. There was only one spot where he could get a grip with his teeth: the head above the spreading hood—and he would have to be faster than he had ever been. The cobras were heading fast for a blot of darkness against the white-painted wall—a hole in the foundation. At Amy's end of the house. Rikki dashed for the hole as the tail of the second snake disappeared, but it was too small for him.

Stupid! He was wasting time! Amy always left the window open a crack so he could get back in. He dashed up the wall and into the window, chattering furiously. Amy was in her bed, and her familiar smell helped to calm him. At first he thought the snakes hadn't been able to fit through the hole. Then a shadow in the far corner uncoiled itself. No time to think: he sprang.

PART 3: MEENA

The sound of her daughter's shrieks woke Meena, who leapt out of bed, ran down the hall and stopped horrified at Amy's door. Thrashing on the floor was the nightmare of her own childhood, a king cobra nearly four feet long—and something else, something furry. Amy's terrified eyes met Meena's over the battlefield. Suddenly the snake's body went flying across the room and lay there motionless. Meena was about to run into the room when Amy screamed again: a second cobra was sliding toward the foot of the bed.

Before Meena could move to get between the snake and her daughter, the little animal leapt for the snake's head as it drew back to strike. He missed, and the snake whipped around to sink its fangs into his back. But withdrawing the fangs slowed it down just enough, and this time, Amy's protector did not miss. He bit down until his teeth met in the cobra's backbone. The snake's body flailed once and was still.

Trembling so hard she could barely move, Meena ran to put her arms around Amy. "Sweetheart, did it bite you? Try to answer me." If Amy had been bitten, a doctor might not get there in time to counteract the snake venom. Amy was sobbing, "Rikki," and stroking the mongoose that lay against her knee. He was bleeding from multiple bites—the first snake must have bitten him too—and he looked up at her weakly. "He's my only friend here," Amy sobbed, "and he tried to save me and now he's going to die!"

"Amy! Stop this immediately and answer me! Did they bite you?"

"No, no, they didn't bite me. Don't take him away!" she shrieked as Meena reached out toward the little animal. "I want him to stay with me!"

Meena sighed. "I'm not taking him away. We'll stay here together." Barathi had come in and screamed too when she saw the dead snakes on the floor. "Barathi, will you make us some tea? Make yourself some, too. And we'll need someone to come and clean those up in the morning." By the time Barathi came back with three cups, Amy had fallen asleep leaning against her mother's side, and Meena stayed up the rest of the night, listening to her daughter breathe—and the thinner, smaller sound of the mongoose breathing too.

He was still breathing when the light came in the window, and his eyes sparkled alertly when she moved. "Amy," she whispered. "Wake up, sweetheart." Amy opened her eyes and saw her friend looking back at her. "Rikki! Mommy, how can he still be alive?"

"When I was growing up, people used to say that a mongoose can kill a snake but a snake can't kill a mongoose. I never knew if it was true or not. But look, Amy, he has bites in five places and he seems fine." She tried to look stern, but she was so relieved that it was hard. "Amy, since when have you been keeping this animal in the house?"

Amy smiled slyly. "You know. A while. But I can keep him now, right?"

Rikki was too tired to move—he was cut, bruised, exhausted from the fight—but he heard Amy's happy tone and snuggled closer to his human. Later, he would have to find the Gray Sisters and tell them that Naga and Nagini were dead, so they and their rat friends should eat themselves sick on those eggs—before they hatched. But a hero deserves a little rest. And a bowl of milk. And maybe some meat off his human's plate. He had a feeling things were about to get even better.

Scientific Connection

Amy is lucky to be alive: the cobra is one of the most dangerous snakes on the planet. Glands in a cobra's head release venom through its fangs when it bites an animal or a human; if a snake bites you, you've been "envenomated" (rather than "poisoned"). Snake venom is typically a cocktail of destructive proteins that wreak havoc in a few different ways. In the case of the cobra the most deadly effect is on the skeletal muscles—all of the muscle tissue that you can move voluntarily. The heart is made of cardiac muscle, not skeletal, so the venom doesn't affect it; nor does it damage brain tissue (so a victim of snakebite is aware of all her or his sensations). Since those are two of the main organs that maintain life, you might wonder why cobra venom is dangerous at all. The inability to move your limbs is unpleasant but not lethal: what kills victims of cobra venom is the paralysis of another skeletal muscle, the diaphragm, whose contractions enable you to breathe. If the diaphragm stops moving, breathing stops as well. You don't have to know much about medicine to know you die if you can't breathe.

There is a protein called the nicotinic acetylcholine receptor whose job is to transmit a message to skeletal muscle tissue to tell it to contract. This protein's job description is like that of a translator: instead of languages, it translates chemical signals from the nervous system into electrical signals that produce muscle movement. When you want a muscle to move, your nerves give a message to the nicotinic acetylcholine receptor, which translates it and passes it on to the muscles, where it makes them contract and move. Any problem with the nicotinic acetylcholine receptor prevents communication between the nerves and the skeletal muscles and results in paralysis.

Many animal venoms are designed to do exactly this in order to disable prey, and the cobra's venom is no exception. One of the proteins in the venom cocktail, alpha-cobratoxin, binds to and disables nicotinic acetylcholine receptors, causing the whole-body paralysis and death described above. The venom may take as long as twelve hours to take effect in a human, but when it does it can easily kill an adult. So how could a much smaller mongoose withstand several bites and still survive? The answer is that the mongoose has alpha-cobratoxin-proof nicotinic acetylcholine receptors.

The secret to these cobra-proof proteins lies in the rough endoplasmic reticulum. One of the purposes of the rough endoplasmic reticulum is to add special modifications to proteins that are being synthesized; usually these modifications enable a protein to carry out a special job. Any protein can be assembled on a ribosome in the cytoplasm, but only the rough endoplasmic reticulum can make special modifications after the protein has been made. One of these modifications attaches a sugar molecule to a protein: this is called glycosylation, and in addition to other functions, it is the key to making the mongoose's nicotinic acetylcholine receptor highly resistant to alpha-cobratoxin. As the nicotinic acetylcholine receptor is synthesized in the rough endoplasmic reticulum of the skeletal muscle cells, it receives glycosylations at several key structural points that prevent alpha-cobratoxin from binding to it and inhibiting it. Though all nicotinic acetylcholine receptors are processed in the rough endoplasmic reticulum, most other animals (humans included) do not receive these essential glycosylations. As a result the mongoose is a natural predator of the cobra while almost every other mammal is its natural victim.

The resistance of the mongoose to cobra venom has been used as a plot point in many tales and astute readers will recognize the above story as an adaptation of the Rudyard Kipling classic "Rikki-Tikki-Tavi". People really do welcome these animals in their homes as a protection against snakes, similar to those who keep cats to catch mice. Talking animals, on the other hand, are a clue that this is a work of fiction; but if some animals could talk, snakes would probably not be among them, as snakes can't hear. Cobras also have extremely small fangs and envenomation really requires a good bite that involves some chewing. The possibility of illustrations of cobras with giant fangs outweighed our herpetological integrity so we hope you can forgive us.

⇀ Take Home Message ⇁

Nicotinic acetylcholine receptors are essential to translating chemical signals from the nervous system to electrical signals that make skeletal muscles contract. Disabling nicotinic acetylcholine receptors causes skeletal muscle paralysis. Paralyzing the diaphragm stops breathing and leads to death. The rough endoplasmic reticulum is involved in performing glycosylations that allow proteins to do special jobs.

THE STORY

Delia was incomprehensible as she sobbed uncontrollably over her husband's corpse. Private Detective Shirley Doyle tried to console the grieving woman: "It's all right. I understand that this is a terribly difficult time for you, Delia. You should get a copy of his Will in the mail later this week from the estate attorney. In the meantime I will call the coroner."

"Why are you even here? I didn't call you. In case you hadn't noticed, I just lost the love of my life. Please just leave me alone. I don't care about a Will...or an estate...or anything. None of it will bring him back!" Delia Sanders, an attractive thirty-something woman, threw herself on her husband's body as her tears soaked the white sheet that covered him. As Detective Doyle walked out of the room, she wondered how anyone could mourn the passing of a vile man like Laurent Sanders whose life had been defined by greed and cruelty.

Laurent had been a very wealthy entrepreneur who owned everything from construction companies to bakeries. He had run the city of Riverside with an iron fist in both official and unofficial capacities. Laurent was a ruthless man whose moral compass didn't have a needle. His corruption was well known. He had been indicted several times for racketeering but both evidence and witnesses always seemed to disappear before the trials. His brother Charles was just as dirty and rumored to be the brains while Laurent was the muscle. Detective Doyle had dedicated her life to bringing the Sanders family down; this devotion had unfortunately gotten her kicked off the force.

"Thank you for coming, Detective Doyle," Charles Sanders said as he put his arm around her and led her into an adjacent room. "Don't believe anything that harpy tells you. The only thing Delia loved about my brother was his money. She will tell you he died in his sleep but I know she killed him. I can't figure out how, though. Laurent would not have gone down without a fight. Even if she had poisoned him he would have strangled her with his last strength. He was only 53! Other than a little cough he used to get he was the picture of health...."

Detective Doyle cut Charles' pleas short by knocking his arm from her shoulder. "Your brother was scum and so are you. All I wanted was to see the both of you rot in a cell as your empire crumbled. As far as I'm concerned, one down, one to go. If it was murder then I can assure you that the list of suspects will not be short."

Charles grew more agitated. "People don't just show up dead like that, not a mark on their bodies, just like they died in their sleep for no reason! He had a cleaner bill of health than I do! He never had so much as high cholesterol. Tell me you don't believe that vapid woman! She's nothing but a liar and a cheat, and I believe that she's responsible for my brother's death! I was the one that called you. We have our issues but you're the best there is and I'll pay any price for justice. I want to hire you to find out how she did it."

"People die in their sleep all the time. Maybe he finally had a fatal attack of conscience..." Detective Doyle paused for a moment as she caught something suspicious out of the corner of her eye. As Delia wailed over her husband's body, Detective Doyle could see a thin stream of blood running down her ankle just beneath the hem of her dress. Detective Doyle turned to face Charles: "You might just be right. $300 a day plus expenses. I'll start my investigation and update you on my findings. I will need full access to the estate. Now leave me to my work."

Detective Doyle kept in the shadows and surreptitiously continued to observe the grieving widow. As soon as she thought everyone had left the room, Delia's sobbing abruptly turned into barely audible laughter. Detective Doyle watched in surprise as Delia spit on Laurent's corpse and danced around the room. Before leaving, Delia closed an air vent beside Laurent's bed and glanced at the carpeted floor frantically, as if to look for a precious item that had gone missing. As soon as Delia left, Detective Doyle emerged from her hiding spot and searched the area around Laurent's bed. Her suspicions deepened as she found a pin with a bloody tip embedded in the carpet underneath the bed.

As Detective Doyle examined the rest of the room she noticed that it did not contain any feminine items. In the closet there were only suits, no dresses. The drawers had only black socks and men's underwear. She found only men's cologne but no perfume. There was a vent beside Laurent's bed that had some strange scaly skin caught in it. Detective Doyle saw a maid cleaning in the hallway and slowly approached her. She was thin and quiet, a slip of a woman. The sound of Detective Doyle's footsteps startled the maid, who dropped her feather duster. "Hello, miss, I'm Detective Doyle. I was wondering if you could tell me how long Laurent and Delia had been sleeping in separate rooms. You have to admit, strange behavior for newlyweds." Detective Doyle shook the maid's hand, deftly slipping a $100 bill into her palm.

As the maid regained her composure it was plain to see a fresh bruise on the side of her face. "They weren't getting along at all. They fought constantly and lived in separate rooms for at least two weeks. Just between us, Laurent said he was going to change his Will this week so she would be cut out. She is such an awful woman. I saw her jab herself in the leg with a pin at least six times before you all came in to see Laurent's body. She started crying after that. When she realized what I'd seen she slapped me to the ground and said if I ever said anything I was next. It's not the first time she hit me either. What else would you expect from a woman raised in a carnival? A cousin of mine in Rabbit Ridge said Delia used to train animals. She trained a monkey to fight people. It's the strangest thing....in her real room there is a big black case and sometimes I swear I hear hissing inside of it. It doesn't matter, now with Laurent dead the whole place is hers and I have a lifetime of working for the devil queen to look forward to. It's this or starve, I guess. My name is Ellie, by the way."

Detective Doyle's eyes lit up as the whole case began to make sense. A wry smile crept across her lips. "Hey Ellie, how would you like to get some payback..."

Two days later Detective Doyle returned to the Sanders estate. She hid in the bushes and waited for the drama to unfold. She didn't have to wait long before Charles' Mercedes Benz screeched into the driveway. Charles smiled as he ran to the front door and began to bang on it vigorously.

Delia answered the door with rage in her eyes, "That's right, you little tramp!" Charles said gleefully to his sister-in-law. "You get nothing, I got the Will in the mail this morning – you inherit everything, after me."

"It can't be true!" Delia whispered, too shocked to yell. "He assured me that I would have everything! How could he do this to me?"

Charles let out a devilish chuckle. "My brother must have seen you for what you are. Now, I'll give you a week to collect your things and get out of my house. Meanwhile, I think I'll spend my first night in the master suite of my new house... you can stay in the guest house."

"Wh-what?" Delia was furious. "You haven't heard the last of me!"

Charles gave a toothy grin "You saw the document – it's legal. Looks like Laurent had you figured out all along. You get nothing as long as I am alive, and I get the entire estate. Pack up and get out!"

Later that night, Detective Doyle knocked on one of the estate's back doors. It opened ever so slightly as Ellie quickly motioned for her to come in. Ellie handed Detective Doyle a broom and the two of them stealthily crept up the stairs toward the master suite. They opened the door just a little and saw Charles fast asleep in bed.

Detective Doyle and Ellie kept their eyes trained on the open vent next to the bed. It was nearly three hours past midnight when it happened. A slender, striped snake emerged slowly from the vent and slithered towards Charles' slumbering body. Detective Doyle kicked open the door and charged towards the snake swinging the broomstick at it. She struck at the snake, but missed and hit the vent with a loud clang. The sound of the impact woke Charles up suddenly. When he saw the snake he panicked and rolled off of the bed to escape it, hitting the floor with a thump. The chaos in the room terrified the animal and sent it slithering back into the vent. "You idiot...why didn't you kill that thing! What are you even doing here?" Charles screamed.

"I didn't want to kill it, just send it back to its master," barked Detective Doyle.

Then they heard a yelp of surprise and terror from the next room. Rushing in, they found Delia, still clutching a small bell, lying next to an open suitcase containing a cylindrical basket where the snake had taken refuge. Detective Doyle looked at Ellie and said, "She's all yours," before tossing her the broom. Delia tried to rush out of the room and escape but her attempt was cut short by a sharp blow to the head courtesy of Ellie's broomstick.

Charles approached Delia's body with apprehension as Detective Doyle put handcuffs on her. "I think she is unconscious. Brilliant, Detective. Really brilliant. But how did you ever guess at such a method of murder?"

"Elementary, my dear Charles," replied Detective Doyle, "Delia's grieving was an elaborate act. She hated Laurent with every fiber of her being. She stood to inherit everything and all that stood in the way were his vital signs. Despite being newlyweds, they weren't even living in the same room. She had to stab herself in the leg with a pin multiple times to be able to generate believable tears. I saw the blood running down her leg and found the bloody pin under the bed. There were bits of a snake skin caught in the coarse metal grate. As soon as I learned that Delia trained animals and she kept a snake as a pet in her room it all made sense. She was afraid Laurent would cut her out of the Will so he had to die before he could act on his threats. Delia trained the snake to return to the basket when she rang a bell...the one we found in her hand. That snake, the many-banded krait, as it's called, is extremely dangerous and has very small fangs. Delia knew that such fangs would produce a wound so fine that a sleeping person would barely notice and no wound would be left. The results of the bite would certainly be fatal. The krait was the ideal silent killer."

"How did you know she would try and kill me tonight" Charles inquired.

"Ha! I slipped fake copies of the Will in both of your mailboxes. I could count on you to be a jerk and make a loud scene and Delia to be vicious and try to kill you. I have the real Will right here. The irony is that neither of you got anything. The whole estate goes to Laurent's first wife Linda whom he recently abandoned for Delia. Apparently he felt guilty and this was his way of making amends. You can give the payment for my services to Ellie, my new assistant. If memory serves me well, you said it would be generous."

Scientific Connection

The snake that bit Laurent Sanders, the many-banded krait (Bungarus multicinctus), produces a lethal venom that is a cocktail of deadly proteins. The most dangerous component of this mixture is α-bungarotoxin. It is a potent neurotoxin that kills the snake's victims by disrupting the neuromuscular junction in skeletal muscles. Skeletal muscles are under voluntary control, which means that when you send a signal down your nerves to tell them to move they do so. The nerves and the skeletal muscles are two separate entities that are not directly connected. The transition point that separates them is known as the neuromuscular junction. Electricity can't cross this gap, but chemicals can. The electrical signal from the nerves is translated into a chemical signal that crosses the gap; when the chemical reaches the muscle it sets off another electrical signal that leads to muscle contraction. When the electrical signal from your brain reaches the end of the nerve, a chemical known as acetylcholine is released from the nerve into the neuromuscular junction. When acetylcholine crosses the gap, it interacts with a protein on the surface of the muscle known as the nicotinic acetylcholine receptor. The nicotinic acetylcholine receptor acts as a translator: it converts the chemical signal of acetylcholine into the electrical signal that causes the muscles to contract. The release of acetylcholine by the nerves and its reception by the muscles is essential to muscle contraction and movement.

The krait venom, α-bungarotoxin, blocks the nicotinic acetylcholine receptors so that they cannot be activated by acetylcholine. The result of this is muscular paralysis. The nerve can send all the signals it wants but the muscles will not move. If a krait bit you, you wouldn't be able to move your fingers, your toes, your arms, or your legs. It would be terrifying to feel your muscles grow weaker and try to move them as hard as you can but to no avail. How does not being able to move any of your limbs kill you? As it turns out there is one skeletal muscle, the diaphragm, that is essential to life. The diaphragm is the muscle that allows you to breathe; if it's paralyzed then you stop breathing. The net result of the venom is respiratory failure and death. Botulinum toxin, which causes the disease known as botulism, has the same effect but instead of preventing acetylcholine from activating the nicotinic acetylcholine receptor it prevents it from being released by the nerves. Astute readers may recognize this as an adaptation of a classic Sherlock Holmes story.

⇢ *Take Home Message* ⇠

The neuromuscular junction is an area of communication between nerves and muscles. The chemical neurotransmitter acetylcholine is responsible for delivering the message from the nerves to the muscles that tells them to move. Any interference with this system results in paralysis and death from respiratory failure.

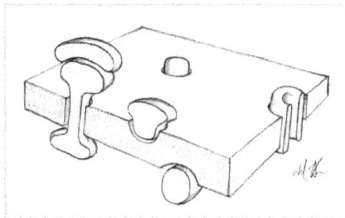

THE STORY

It was summer in the small rural town of Rabbit Ridge and the first night of the County Fair. Despite the tractor pull, the Tilt-a-Whirl and the fried-dough stand, the Jenkins brothers were confident that they'd be the biggest attractions, at least to all the good-looking local ladies. Working on the family dairy farm had transformed them from skinny kids to bulky men whose every move showed in a ripple of muscle. Dave, the oldest, was 6 foot 4 and 280 pounds of rock hard muscle, and his brothers Art and Cliff were nearly as massive. He claimed to be the strongest man in the county; he liked to uphold his reputation by bending steel nails in his fingers and crushing beer cans before they'd been opened. He knew the Fair would offer him and his brothers plenty of chances to show off. They showered off the day's sweat, slapped on some Old Spice cologne, and piled into Dave's Dodge Charger.

First stop, the High Striker: "Step right up gentlemen show your ladies what a strong fella you are take a hold of that hammer now ringing the bell at the top wins the big pink bunny!" the barker chanted in one breath. Nelson Lester, the town clerk still wearing his bow-tie from work, had just managed to lift the hammer two-handed and was about to swing it when Dave Jenkins shoved him aside and snatched the hammer from him on his way down. While Nelson struggled to recover his dignity and his footing, Dave pulled back the hammer, lifted it over his head easily with one hand, and smashed it down on the platform. The puck shot to the top, rang the bell, and stuck there. "You miserable—" the barker squawked, "you broke my concession, you hick meathead!"

"Who you callin' names, skinny?" Dave sneered. "This machine might be tough for the weak-kneed fools from Riverside, but it's not my fault it can't stand up to a real man. Now back off, or the next thing I break might be you." His brothers slapped him on the back and laughed. All three of them continued to ignore Nelson as he brushed himself off and disappeared behind the Whack-A-Mole. He reappeared in a few minutes walking quietly behind a stocky, bow-legged man chewing on an unlit cigarette. "Hey, boys," said the stranger, "I hear you're supposed to be pretty strong. You don't wanna waste your time on this rinky-dink piece of equipment, you

really wanna show off what you can do. Am I right?" When the brothers nodded eagerly, the stranger said, "Well, what I recommend you do is wrestle my friend Pinky. Pinky weighs about a hundred and fifty pounds, and I bet you a thousand to one you can't wrestle him to the ground." As the brothers coughed and choked in fake astonishment, the stranger went on, "I got three thousand dollars right here that says you can't. Whaddya say, boys, one dollar from each of you, a thousand to one, three thousand to three?"

By this time a little crowd had gathered around what was left of the High Striker, and in it Dave couldn't help but noticing a group of pretty young women. "Yeah, okay," he said loudly, "show me where Pinky's at. Anybody else wanna come along and see the show?" The crowd—including the girls—trailed after the stranger and the Jenkins brothers to a giant cage set back from the midway. Around it was a much bigger crowd of people, and in it was a chimpanzee whose head was about as high as Dave's Harley-Davidson belt buckle. It seemed peaceful and calm, ignoring the hooting and hollering of the onlookers. "Now all you gotta do," the stranger was saying, "is pin Pinky once in five minutes or less. The only rules are you gotta wrestle, not punch. You punch him, I can't be responsible. Pin him in five or less and the money's yours."

The brothers went into a huddle and decided that Dave should go first in case there was some kind of catch — but how hard could it be to beat a little monkey like that? Dave swaggered through the cage door and then flinched as a handful of chimp feces hit him right in the center of his best muscle shirt. The crowd shrieked with glee and Pinky's teeth showed. Dave roared, "I'm gonna tear you up, you pipsqueak!" and charged. Next thing he knew, his ribs were aching from where he had hit the cage bars. His brothers were shouting, "Dave, get up, Dave, come on, bro, don't let us down! You still have three minutes!" Dave struggled to his feet, screamed a few curse words, and flung himself toward Pinky, who stood in the center of the cage looking bored. With one long arm, he slammed Dave to the floor, dragged him up and threw him again, over and over while the crowd cheered and Dave cursed the day he was born. Feeling a hand grip his arm again, he shouted, "Get your dirty no-good paws off me, you blasted ape!" before realizing it was Pinky's owner, pulling him to safety and handing him a block of ice in a dirty towel for his jaw.

"Where are my brothers?" Dave said through his swollen mouth.

"Coupla big fellas like you? They ran off when they saw Pinky had the drop on you. I wouldn't feel too bad, son. Pinky's wrestled about five hundred guys and beat about five hundred guys."

"That's the last three dollars you'll ever make off that monkey," Dave warned. "Nobody's gonna pay to wrestle it after seeing what it did to me."

"Nah, see, you got it all wrong," Pinky's owner told him calmly. "What you paid was an entry fee. Everybody else paid an admission fee. Not too many people will pay to wrestle an ape, but a lotta people will sure pay to see a big knucklehead get beat. Now if you want my advice..."

Dave told him what he could do with his advice. "Suit yourself," said the stranger calmly. "By the way, if you're wondering what happened to all those good-looking women, most of 'em dispersed and a couple of 'em walked off with a little fella in a bow tie. I'll see you around, I guess."

"Not if I see you first," Dave growled, but Pinky's owner had already disappeared into the fairground crowd. There was nothing left for Dave to do but start the long, painful hobble toward the exit.

Scientific Connection

All the best parts of this story really happened: long before the days of the S.P.C.A, ape wrestling was a common "attraction" at carnivals throughout western Pennsylvania and West Virginia, and the humans always lost. It is definitely illegal now. Forcing animals to fight and keeping them in horrible living conditions is animal abuse—much like Michael Vick's conviction for dog fighting. However, unlike many abused animals, Pinky was able to take some revenge on those who mistreated him.

Pound for pound, apes are significantly stronger than humans. Even a very large and strong man, like Dave, would have a hard time in a contest of strength with a chimpanzee. You can see evidence of this physical disparity in this clip where a 180-pound orangutan destroys a sumo wrestler in a game of tug of war without much effort: http://www. metacafe.com/watch/60685/sumo_wrestler_vs_female_orangutan/. How can apes be so much stronger than humans when they are often physically much smaller and lighter? One hypothesis claims it's because of their differing muscle physiology, specifically the size of motor units and their recruitment.

A nerve and the skeletal muscle cells that it controls are known as a motor unit. If a nerve controls a lot of muscle cells it is considered large, and if it controls very few it is considered small. The more muscle cells in a unit, the more force it will be able to exert. Large motor units are often used for brute strength activities that require a lot of force but little precision (like Dave's beer-can-crushing party trick) while small motor units are used for activities that require more control but less force (like playing a piano).

"Recruitment" in motor units refers to the application of force. The amount of force applied in a task depends on how many motor units you recruit to perform it. Touching your nose with your index finger is the same motion as curling a heavy dumbbell, but you don't use the same amount of force for both tasks. A motor unit is either "on" and contracting with full force, or "off" and not contracting at all: curling the heavy dumbbell would require a lot of motor units to be recruited while touching your nose would require far fewer.

Humans are born with a fixed number of skeletal muscle cells, and these cells don't normally undergo mitosis, which means that the muscle cells you are born with are all you will ever have. This sets the upper limit for your strength. Why did Dave lift the hammer for the High Striker so easily while Nelson struggled with it, even though they likely had the same number of muscle fibers and could recruit similar numbers of motor units? The strength training that Dave did by working on the farm made his muscle cells larger and capable of producing more force than Nelson's (this is known as hypertrophy, and you can see it in bodybuilders) but did not increase their number—no amount of exercise would do that.

So why couldn't he win against the chimpanzee? Pinky was able to physically dominate Dave—and could easily have ripped him apart—because a much higher proportion of an ape's skeletal muscle system is devoted to large motor units. An ape can therefore recruit more large motor units and get more force out of its muscles than humans can. The trade-off is that an ape has fewer smaller motor units left for fine muscle control. While there will never be any great ape pianists, they will crush a human in a physical contest any day of the week.

⇀ *Take Home Message* ↽
A nerve and the muscle fibers that it controls are known as a motor unit. Multiple motor units can be recruited to complete tasks and the more that are recruited the more force will be exerted.

Further Reading

1. *Use this link to read more about the motor unit hypothesis:* http://www.livescience.com/5370-chimps-stronger-humans.html

2. *This is an eerily similar story that follows the same general plot, though it was an independent event. It has all the key points: big guys; a carnival; prospect of wrestling an ape; and things not ending well:* http://peterlumetta.hubpages.com/hub/HOW-NOT-TO-WRESTLE-AN-APE *Index of Concepts*

Authors and Illustrators

Shunned by society, the cellular organelles fulfill their solemn vow to ensure our continued survival. Cellular organelles are the untold heroes of the biological world. They work tirelessly, never leaving the inside of your cells, yet they remain neglected and underappreciated by those whose very lives depend on their toil. If they didn't work hard all day, life would not exist. Sure there is "Boxing Day", "Guy Faux Day" and even "Groundhog Day" but have you ever heard of "Ribosome day", "Endoplasmic Reticulum Day", or "Golgi Apparatus Day?" Do you even know what the Golgi Apparatus does? Well, it knows what you do. If organelles didn't exist then there would certainly be no boxing, no Guy Faux (which may have been a good thing but that is beside the point), and certainly no groundhogs (definitely bad). Well, rest assured all you Cell membranes, nuclei, and peroxisomes.....Today is your day to shine because the Providence Alliance of Clinical Educators presents a science book like no other.